Sensory Neural Networks: Lateral Inhibition

Sensory Neural Networks: Lateral Inhibition

Bahram Nabet

*Department of Electrical and Computer Engineering
Drexel University*

Robert B. Pinter

*Departments of Electrical Engineering and Zoology
University of Washington*

CRC Press
Taylor & Francis Group
Boca Raton London New York

CRC Press is an imprint of the
Taylor & Francis Group, an **informa** business

First published 1991 by CRC Press
Taylor & Francis Group
6000 Broken Sound Parkway NW, Suite 300
Boca Raton, FL 33487-2742

Reissued 2018 by CRC Press

A Library of Congress record exists under LC control number: 91015238

Publisher's Note
The publisher has gone to great lengths to ensure the quality of this reprint but points out that some imperfections in the original copies may be apparent.

Disclaimer
The publisher has made every effort to trace copyright holders and welcomes correspondence from those they have been unable to contact.

ISBN 13: 978-1-138-10546-1 (hbk)
ISBN 13: 978-1-138-56181-6 (pbk)
ISBN 13: 978-0-203-71041-8 (ebk)

Visit the Taylor & Francis Web site at http://www.taylorandfrancis.com and the CRC Press Web site at http://www.crcpress.com

Cover Design: Chris Pearl
In-Text Graphics: Falaq Pulliam, ETP
Copy Editor: Anthony Lafrenz, ETP
Composition: Electronic Technical Publishing Services Company

To you, Behrooz.

B. N.

To Marie.

R. B. P.

Contributors

R. J. BELSHAW is with the Department of Engineering System Design at the University of Waterloo.

ABDESSELAM BOUZERDOUM is with the Department of Electrical Engineering at the University of Washington.

ROBERT B. DARLING is with the Department of Electrical Engineering at the University of Washington.

M. E. JERNIGAN is with the Department of Engineering System Design at the University of Waterloo.

G. F. MCLEAN is with the Department of Engineering System Design at the University of Waterloo.

BAHRAM NABET is with the Department of Electrical and Computer Engineering at Drexel University.

ROBERT B. PINTER is with the Departments of Electrical Engineering and Zoology at the University of Washington.

Contents

Abstract

Sensory information is detected and transformed by sensory neural networks before reaching higher levels of processing. These networks need to perform significant processing tasks while being compatible with the following levels. Lateral inhibition is a mechanism of local neuronal interaction that gives rise to significant global properties. This book studies visual sensory neural networks whose activity is governed by nonlinear lateral inhibition. It studies biological bases of models of lateral inhibition, computational properties of these models stressing their short term adaptive behavior, their relation to recent activity in neural networks and connectionist systems, their use for image processing applications, and their application to motion detection. Analog hardware implementation of these classes of networks are described in different technologies and results of implementation which corroborate theoretical analysis and show technologically desirable applications are presented.

Finally, nonlinear mathematical techniques are used to analyze temporal and spatial behavior of these models with the latter showing high order classification properties of the networks. As an interdisciplinary work, this book provides a consistent but multifaceted view which is useful for neural network theorists, biologists, circuit designers, and vision scientists.

1

Introduction

This work is part of a general research effort whose final goals were aptly described by one of the most eloquent researchers of the field, Warren S. McCulloch (1961),[1] as

> "The inquiry into the physiological substrate of knowledge which is here to stay until it is solved thoroughly, that is until we have a satisfactory explanation of how we know what we know stated in terms of the physics and chemistry, the anatomy and physiology, of the biological system."

In pursuing such an aim, Seymour Pappert (McCulloch, 1965) cautioned that:

> "We must, so to speak, maintain a dialectical balance between evading the problem of knowledge by declaring that it is *"nothing but"* an affair of simple neurons, without postulating *"anything but"* neurons in the brain."

This manuscript has the modest goal of presenting a *coherent* view of one of the simplest, and most fundamental, mechanisms that connect simple neurons into *networks*. In doing so Pappert's dictum was adhered to and, it is hoped, a contribution made to the field of neural networks as defined by McCulloch.

As a reflection of the field it is describing, this work has to make several *connections:*

- It has to be related to what has re-evolved into the *field of neural networks*, keeping both its new computational and biological trends in perspective.

[1] This paper has the intriguing title: "What is a number, that a man may know it, and a man, that he may know a number?"

- It has to be based on a solid mathematical foundation, and contribute to the understanding of these networks.

- It has to keep abreast of technological advances in materials, devices, and circuits in general and progress in the implementations of neural networks in particular.

This book is organized as a merger of these aspects. Chapter 2 describes the biological bases and historical background of the network of neurons related by lateral inhibition. Chapter 3 relates the described models to recent efforts in neural network modeling, stressing the properties that are suitable for processing of sensory visual information. Chapter 4 reports on the application of models of nonlinear lateral inhibition to image processing. Chapter 5 studies preferential directional and motion selectivity properties of these networks.

Chapter 6 discusses a general framework for electronic, and optoelectronic implementation of nonlinear lateral inhibition. This topic is further elaborated in the next chapter where choice of different technologies is detailed. Chapter 8 reports the results of actual electronic realization of a prototype model that although very simple, corroborates the theoretical results of the previous chapters. The implementation results prove that networks of nonlinear lateral inhibition, and implementations with *few* transistors, are indeed capable of capturing some of the most salient features of peripheral vision which also have direct technological application. Among these are edge enhancement, dynamic range compression, feature extraction, adaptation to mean input intensity, tuning of the receptive field and modulation transfer function, tunability of the sensitivity, directional selectivity, and coding of the intensity.

Chapter 9 explores the mathematical "connection" by first describing the stability of the network and then using the Volterra-Wiener series expansion to determine its temporal and steady-state spatial behavior. The results of spatial expansion are shown to be directly related to classification properties of the network.

Given the plurality of aspects, each chapter is written to be self-consistent and independent, hence some introductory parts of the chapters may slightly overlap, but this was deemed necessary for readers of different backgrounds and interests. Each chapter also includes a "discussion," or "conclusions" section that highlights the important results and establishes the continuity of the topics.

Preparation of this manuscript, of course, far exceeded the initial estimates of the required time and effort. This work would not have been completed without the support of Zohreh Nabet and Marie Harrington. The work of H. K. Hartline and F. Ratliff encouraged a fascination of biological visual systems for R. B. Pinter which now spans three decades; this book is also a tribute to Floyd Ratliff. The authors have benefited from numerous discussions with colleagues and are especially indebted to those who agreed to contribute to this volume for a broader and more detailed coverage. The wit, wisdom, and encouragements of Professor David L. Johnson are greatly appreciated. Mr. Russ Hall of CRC Press suggested the publication of this monograph and patiently supported and guided its preperation. This work was supported by the National Science Foundation through grants MIP8822121 and BNS8510888. B. Nabet gratefully acknowledges the support provided by Drexel University's Research Scholar Award.

2

Biological Bases

2.1 A BRIEF HISTORY OF LATERAL INHIBITION

The concept of lateral inhibition arose in the extensive experimental research of H. K. Hartline and colleagues on the facetted compound eye of the familiar "Horseshoe crab" *Limulus* (Xiphosura). This research occupied a period of over fifty years and is an outstanding example of bringing quantitative mathematical methods of signal transmission to bear on a biological preparation (Hartline and Ratliff, 1974). The *Limulus* is an animal which appears not to have evolved further from its form in the time of the trilobites (Middle Paleozoic era). This animal is thus among the most primitive living. Yet, we can most easily describe the operations of lateral inhibition in mathematical terms. It is a quantitative, precise mechanism which was synthesized by evolution of this organism, very early in geologic time. Lateral inhibition is simultaneously a biological principle and a mathematical description of a biological neural network. Currently an electronic synthesis of lateral inhibition would be termed a sensory neural network.

The facets of the compound eye of *Limulus* are the largest found, approximately 100 micrometers (microns) in diameter, and the photoreceptor cells are quite accessible to electrophysiological-experiment methods. The structure behind (proximal to) the clear facets, and which includes them as the crystalline cones, is termed the ommatidium (from the Greek for "little-eye"). The ommatidium contains the primary photoreceptor cells termed retinular cells; and the secondary cell, the eccentric cell, which sends nerve impulses (spikes) to the brain of the animal along an axon in the optic nerve. The retinular cells are arranged about the central axis (dendrite) of the eccentric cell in the

3

manner of the slices in an orange. The visual pigment (rhodopsin) is within the retinular cells, which signal light via a slow, or generator, potential which is communicated to the eccentric cells. Proximal to the retina and its layer of ommatidia is the neural plexus, which contains the cross-connections among the eccentric cells that mediate the lateral inhibition. These connections lie in several levels, each of a different dominant neighbor extent, and have been extensively documented by Fahrenbach (1985). From visual neurophysiological experimental analyses, the dominant lateral inhibition is very weak beyond approximately eight facets, having a maximum effect at the third facet or neighbor. This is true for the horizontal (anterior-posterior axis) and scaled somewhat smaller for the vertical (dorsal-ventral axis). From such considerations the Hartline-Ratliff equations have been synthesized, and these are:

$$r_i = e_i(I_i) - \sum_{j=1}^{n} k_{pj} \left(r_j - r_{pj}^0 \right), \tag{2.1}$$

where

$$k_{pj} = 0 \text{ if } \left(r_j - r_{pj}^0 \right) < 0.$$

This is a system of linear algebraic equations when each response r_i is positive and above threshold. The response is the spike firing rate of the ith eccentric cell, which receives excitation e_i from the retinular cells as a result of generator or slow potential responses to the facet's incident light I_i. The transformation $e_i(I_i)$ is nonlinear both in steady-state and time-dependent dynamics (Fuortes and Hodgkin, 1964; Pinter, 1966): the output is a compressed version of the input, similar to a logarithmic relationship, often designated a Weber-Fechner law. The time dynamics have a leftward pole migration as a function of the mean input light intensity or flux (Pinter, 1966). The Hartline-Ratliff equations (2.1) are feedback or recurrent and can often be approximated by a small number of iterated levels of feedforward or nonrecurrent equations (Varju 1962; for a concise development see Ch. 3, Ratliff 1965).

Because of the nonlinear transformation $e_i(I_i)$ and the thresholds, these equations are not linear, but often the experimental parameter set or theoretical analysis is operating in the linear range, where all responses are suprathreshold and the light flux variations in time and space are of low contrast at some given mean level I_0. In this case (2.1) are simply spatially discrete linear filters. The coefficient sets of the k_{pj} generally do not depend on the absolute values p, j but only on the difference function over space, $|p-j|$, and approach zero for $|p - j| > 10$ (Finite impulse response discrete spatial filter "FIR") with a maximum at $|p - j| = 5$. The coefficient set can be approximated with a continuous function which is the spatial impulse response of the system (2.1) (Ratliff et al. 1969).

For spatial impulse response functions possessing at least one maximum, the system (2.1) processes the input e_i across spatial dimension i into an output r_i. The discontinuities and near-discontinuities in e_i are accentuated in response space r_i by the appearance of overshoots and undershoots, or ringing, which is termed the "Mach band." Originally the Mach bands referred to the visual perception of darkening near the dim side and lightening near the bright side of a gradual or ramped edge. This was discovered in a

long series of perceptual experiments by the physicist Ernst Mach and described in a
series of papers on the interdependence of retinal points (1865–1906), available in En-
glish translation by Ratliff (1965). However, the perceptual phenomena are altered for a
discontinuous step (sharp edge) such that another "edge detector" selective for that step
discontinuity predominates or supplants the Mach band perception (Ratliff 1984). Mach
sought to explain these perceptual "illusions" in the function of a neural network that
was known, anatomically, to exist in the retina and the brain. In the early studies he
proposed a model for the response r to the luminance distribution $I(x)$:

$$r = a \log \left\{ \frac{I(x)}{b} \pm k \frac{d^2 I(x)/dx^2}{I(x)} \right\}. \tag{2.2}$$

There are suitable upper and lower bounds on the value of $I(x)$, such that $I(x)$ is
not zero and its range of values is limited, relative to two decades. The positive sign
applies for a negative second derivative of $I(x)$, and a, b and k are constants. In the
later studies Mach proposed a reciprocal inhibition model which resembles shunting or
multiplicative lateral inhibition:

$$r_p = I_p \frac{I_p \sum_j \Phi(x_{jp}) \Delta a_j}{\sum_j I_j \Phi(x_{jp}) \Delta a_j}. \tag{2.3}$$

The inhibition function $\Phi(x_{jp})$ is a positive, monotonic decreasing function of the
distance $|p-j|$ between retinal points.

The most obvious effect of the lateral inhibition operations (2.1), (2.2), or (2.3)
is to produce a response of enhanced contrast relative to the input. At regions of high
contrast this result takes the form of Mach bands. The accentuation of edges, which
is the differentiating nature of lateral inhibition can be considered a deblurring operation, an attempt to restore contrast information lost by blurring. Blurring is necessarily
the physical result of the non-infinitesmal width of the central maximum of the spatial
impulse response. Another function of lateral inhibition may be to reduce redundancy.
The inhibitory connections collect information from a wider area than the excitatory,
so that the inhibitory receptive field region may be viewed as an estimator of a central
local luminance level, and only deviation from that estimation, as signalled by a difference between excitation and inhibition, is transmitted. This hypothesis leads to a quite
interesting theory of adaptation of form of receptive fields (Srinivasan, Laughlin, and
Dubs 1982). Thus a further concomitant function is likely to be the limitation of the
dynamic range which must be utilized by the nervous system in signalling luminance
distribution, and functions thereof, from the visual space. Since the variation, or contrast, of the luminance distribution often carries the salient information for an organism,
only the difference from the mean need be transmitted. This difference distribution has
a more limited dynamic range than the luminance distribution (Laughlin 1989). While
these effects, and the neural network Equations (2.1), (2.2), and (2.3) may appear on
first sight as simple and straightforward, the overall visual impact in simulations of the
lateral inhibition processing can be complex and dramatic (see for example, Ratliff 1965;
Stockham 1972; Jernigan et al. 1989; Belshaw, this volume).

It is not only the visual system which contains clearly proven lateral inhibition, but also tactile and auditory systems (Bekesy 1967; Ratliff 1965; Moller 1987). The function of lateral inhibition may exist well beyond sensory systems, and into the central nervous system. Its functions may be not only those discussed above, but also a means of an organism's synthesis of nonlinear adaptive filtering of information from the sensory system.

2.2 CENTER-SURROUND ORGANIZATION AND RECEPTIVE FIELDS

The receptive field of a visual cell is defined as the region of visual space over which any response of the cell is obtained. The visual stimuli used to make this observation initially would be points or bars, of positive (brightening) or negative (dimming) contrast. It is often the case that there are antagonistic responses due to a given visual stimulus presented in the surround of the receptive field as opposed to the center of the receptive field. A pattern of organization of inhibitory surround concentric with or adjacent to the excitatory center, or region, resembles the lateral inhibition discussed above. The measures of cell response are several: various aspects of the often complicated response of the cell, for example, the peak response, the time-integrated area of response, or short-term or long-term steady state response.

The linear lateral inhibition discussed is a statement of connections, and requires a solution (by, e.g., matrix inversion; see Chapter 9) to define a receptive field. The linear receptive field is the output of the discrete spatial lateral inhibitory filter for a Dirac delta function (impulse function) visual space input. The linear receptive field is the spatial impulse response, weighting function or first order kernel in a Volterra or Wiener series (see Chapter 9). In Marr's classic treatment (1982) the $\nabla^2 G$ operator is the receptive field. However, when the lateral inhibition or the receptive field are not linear (e.g., the receptive fields of retinal ganglion Y-cells, or cortical complex cells) there is no known, closed form transformation from the lateral inhibition connection scheme to the receptive field, and vice versa (Pinter 1987a; Pinter and Nabet 1990). Only the linearization process will allow applications of matrix algebra and linear systems theory to the transformation of the lateral inhibition to the receptive field and vice versa. Furthermore, beyond the purely spatial considerations there is often great complexity in the temporal relationships of higher order visual interneurons, far beyond that found for *Limulus* compound eye eccentric cells. Useful descriptions of the activity of the cell then require a complete spatio-temporal analysis (Yasui et al. 1979; Curlander and Marmarelis, 1987).

An early quantitative experimental and theoretical analysis of cat retinal ganglion cells by Enroth-Cugell and Robson (1966) demonstrates clearly such complexities in the spatio-temporal response for the retinal X- and Y-type cells. That the often assumed linearity of the X-type cell is not precise can clearly be seen along with the decidedly stronger rectification properties of the Y-type cells (Enroth-Cugell and Robson 1966). There are many very nonlinear retinal ganglion cells (Troy et al. 1989). Motion analysis by cortical visual interneurons requires strong nonlinearity in spatio-temporal receptive fields (Emerson et al. 1987; Chen et al. 1989).

It is the thesis of this book that the nonlinear aspects of vision are the most interesting, useful for visual analysis and potentially the most productive for engineering visual systems design.

2.3 ADAPTIVE LATERAL INHIBITION

A long tradition of psychophysical investigation of linear filtering of images by the human visual system was established following the pioneering work of Ernst Mach (Ratliff 1965). Some of this approach is reflected in the linear operators and analysis discussed by David Marr in his book *Vision* (1982) and by Martin Levine in his book *Vision in Man and Machine* (1985). However, it is clear that there are significant nonlinearities and adaptive filtering in the visual system.

The perceptual spatio-temporal frequency response, known also as a Contrast Sensitivity Function (CSF), is obtained by subjects' adjustment or choice of threshold contrast at a set of points on the frequency plane. This is a linear model of the filtering properties of the visual system. However, it is found that as the mean luminance level of the stimulus is raised, the frequency response changes from low pass to band-pass, the upper band-limit increases, and the low frequency gain decreases. What are the causes of this distinctive adaptation? There is of course the primary effect of the lower information capacity at lower luminance levels, because there are fewer photon absorption events to signal the different parts of a given visual pattern. This might be evident as shot noise, or lower relative signal levels. Yet we do not see a pointillist or patchy pattern as luminance is lowered. There may be many reasons for this, but the visual system is without question low pass and low bandwidth at low luminance levels. (This is discussed further in Section 6 below.) Adaptive nonlinear filtering, via nonlinear lateral inhibitory networks and receptive fields may be another important mechanism in these adaptive visual changes (Pinter 1985). Preceding this primarily spatial adaptive filtering, it is well known that the photoreceptors themselves adapt temporally such that they become significantly slower at low luminance (Pinter 1966; Wong and Knight 1980).

Conversely, at higher luminance levels, visual systems become stronger differentiators. This adaptive differentiation was best described for human visual perception by Kelly (1975) in a study where the perceptual experiment gave a family of contrast sensitivity functions each taken at a different mean luminance (see Fig. 2.1). Thus for a given mean luminance, a model of the contrast sensitivity function is a linear Modulation Transfer Function (MTF) followed by a fixed level threshold. To obtain a prediction of the receptive field in this case, one then takes the inverse Fourier transform of the MTF with the assumption that the MTF is a real, even function. The predictive power of this receptive field is substantial (Kelly 1975). A key point in Fig. 2.1 is the adaptation of the MTFs to mean luminance levels. Since the spatial differentiation in the spatial domain is given as ∇^2, this accounts for the decrease of the MTF at low frequencies. The mean luminance controls this spatial differentiation. The model for the adaptive control of spatial differentiation adopted by Kelly (1975) is to replace the ∇^2 operator with $(1 - a(L_0)\nabla^2)$, where $a(L_0)$ is a monotonic increasing function of mean luminance L_0. Lateral inhibition in this context can be shown to be related to this adaptive differentiation simply by

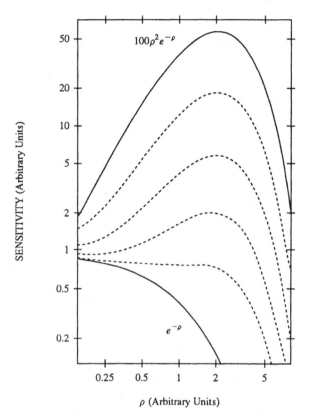

Figure 2.1   Theoretical effect of varying the amounts of spatial differentiation on the sine-wave contrast sensitivity (MTF) as a function of spatial frequency ρ of a single ganglion cell. The lower solid and dashed curves represent the function $(1 + a\rho^2 e^{-\rho}$, for $a = 0, 1, 3, 10$ and 30. The upper solid curve shows the standard contrast sensitivity (MTF) derived from a fit to human subject data. Reprinted with permission from *Vision Research*, 15, D. H. Kelly; Spatial frequency selectivity in the retina. 1975. Pergamon Press PLC.

writing the one-dimensional discrete version of the operator $(1 - a(L_0)\nabla^2)$ on blurred luminance $G(x)$, which is

$$1 + [a(L_0)(-G(x_{i+1}) + 2G(x_i) - G(x_{i-1}))/(x_{i+1} - x_i)^2]. \qquad (2.4)$$

This constitutes a nearest neighbor lateral inhibition in that the central term is diminished or inhibited by the nearest neighboring terms. It is adaptive second differentiation, or adaptive nearest neighbor lateral inhibition. For the MTF, this differentiation causes the low frequency information to be suppressed, more severely as mean luminance increases. Simultaneously there is an increase of high frequency response (Fig. 2.1). The luminance $L(x)$ is blurred by the blurring function $g(x)$ such as that found by Kelly

(1975) as $(r^2 + 1)^{-3/2}$. Often the Gaussian blurring function advocated by Marr (1982) is used, in a convolution operation $g(x) * L(x) = G(x)$. Blurring or smoothing is an inescapable result of physical aperture and sampling effects in physiological optics and physiology of the visual system. Further, there is a clear physiological motivation for a nonlinear lateral inhibition, either post- or pre-synaptic, which possesses the adaptive differentiation property, and mimics the behavior of the differentiation of Eq. (2.4).

There are numerous other examples of luminance adaptation of the MTF in human vision (Patel 1966; Daitch and Green 1969; Fiorentini and Maffei 1973) and in electrophysiological studies of visual interneurons. The adaptive change resembles that of Fig. 2.1. The adaptive change of MTF was shown for cat retinal ganglion cells (x-type) by Enroth-Cugell and Robson (1966). In Fig. 2.2, for a series of mean luminances and temporal frequencies, the spatial frequency response of an approximately linear x-type cat retinal ganglion cell is shown, taken from the work of Derrington and Lennie (1982). In the optic lobe of the fly, the H1 DSMD movement detector neuron, a third order cell in the medulla, adapts similarly over a range of four to five decades of mean luminance (Srinivasan and Dvorak 1980). Carp (Toyoda 1974) and catfish retinal interneurons show similar luminance adaptation not only in their linear properties, the MTF and convolution or first order kernel, but also in their nonlinear properties given by second order Wiener kernels. As mean luminance increases, the center of gravity of the second order kernels moves toward the origin (Sakuranaga and Naka 1985).

2.4 EVIDENCE FOR MULTIPLICATIVE NONLINEAR LATERAL INHIBITION

Experiments in vision involving gain, frequency response and convolution kernels provide evidence for parametric changes of lateral inhibition controlled by mean luminance. The inhibition increases and in some cases narrows toward the excitatory center as mean luminance is increased. For nearest neighbor lateral inhibition (as in Eq. 2.4) this has the effect of increasing the amount of spatial differentiation as mean luminance increases. This also relates to the greater available information in a given luminance pattern at higher luminance levels arising from the greater rate of photon absorption events. (This is discussed in Section 2.6.)

A specific nonlinearity found in several receptive fields of visual interneurons is a *multiplicative* interaction between interdependent points in the field. First, the nature of this will be described, and then it will be related to the parametric changes in lateral inhibition described earlier.

A much studied visual interneuron is found in the ventral nerve cord of acrididae, especially in grasshoppers and locusts. This unit, termed Descending Contralateral Movement Detector (DCMD) has the largest diameter axon in the ventral nerve cord and thus its response to visual patterns is easily accessible to simple methods such as silver wire hook electrodes and conventional amplifier. It is likely that the DCMD is involved in a visually triggered escaping jump response, and collision avoidance in flight (Reye 1990). In response to a small (3 visual degrees) contrasting point of the visual field which either moves, or dims, or brightens, the DCMD responds with a train of

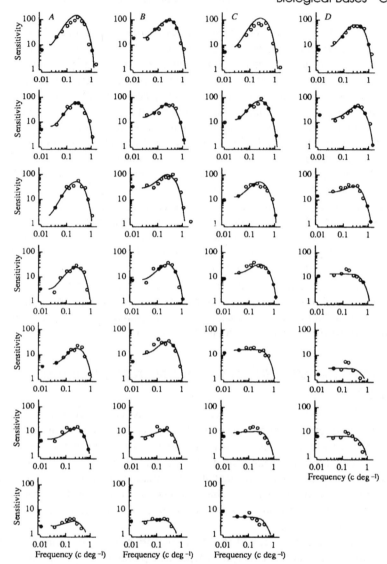

Figure 2.2 Spatial contrast sensitivity functions for an X retinal ganglion cell (25J) measured at four temporal frequencies (columns, from left to right: *A*, 0.65 Hz; *B*, 2.6 Hz; *C*, 10.4 and *D* 20.8 Hz) and seven levels of mean luminance (rows, from top to bottom: 200, 15, 1.1, 0.087, 6.6×10^{-3}, 4.9×10^{-4}, 3.8×10^{-5} cd m^{-2}). Filled points on ordinates are sensitivities to flickering uniform fields. Smooth curves fitted to the points are derived from difference of two Gaussian functions. Reprinted from *J. Physiol. 333*, p. 359, "The Influence of temporal frequency and adaptation level on receptive field organization of retinal ganglion cells in cat; 1982. By permission of the Physiological Society and the authors.

spikes of duration a few hundred milliseconds, decreasing in frequency from the initial peak response. From this type of experiment a receptive field can be mapped, which is nearly a hemisphere, and most sensitive (in terms of numbers of spikes in response to the movement from a starting position) along a horizontal line, toward the posterior (caudal) of the animal (Palka 1967; Fraser-Rowell et al. 1977; Pinter 1977, 1979). This region is one from which potential predators may approach the animal.

The receptive field for response to one stimulus as described above does not resemble a linear spatial superposition when two stimuli are present in the receptive field. Despite the fact that to one stimulus the response is excitatory, there is lateral inhibition embedded within this receptive field, and it can be revealed in the following ways. First, if the small visual object eliciting response is made larger in area, including along the axis perpendicular to movement, its excitatory effect is diminished. Second, the excitatory response is diminished by simultaneously moving a second small visual object in the field, anywhere within a space of approximately 50 degrees visual angle. Third, nearby (within 50 degrees) gratings of mid- to high-spatial frequency range can cause inhibition of the small object motion response (Pinter 1979; approximately 0.05 to 0.5 cycles/degree, radial gratings). Here, the main point of these inhibitory effects is that interaction of the two or more stimuli in the receptive field is not predicted by any linear superposition of the response to one point stimulus. The receptive field for one point stimulus does not resemble the receptive field for the interactions in the response to two or more stimuli. A parallel more quantitative development of the lack of linear superposition, and nonlinear interactions within the receptive fields of cat cortical complex cells has been carried out by Movshon et al. (1978) and Emerson et al. (1987).

There are several methodologies to elicit nonlinear behavior of these receptive fields. The most general is to apply a band-limited white noise spatio-temporal stimulus and apply the Lee-Schetzen cross-correlation algorithm to this input and the output to obtain the linear convolution kernel (first order) and higher order kernels of the Wiener series describing the system (Marmarelis and Marmarelis 1978). The second order kernel often contains the major part of the nonlinear operators' effects, and can be visualized as two linear operators each feeding a single multiplier or a set of such three component blocks in parallel (Schetzen 1980). Other methods are to apply combinations of dots or bars along the receptive field two at a time (Emerson et al. 1987) or M-sequences (Sutter 1987). Often the result of these experimental analyses contains a model whose significant nonlinearity is a multiplier, where the response of the interneuron depends on the product of activities in two portions of the receptive field. On the other hand, from a purely mathematical point of view, a nonlinear receptive field function can also be approximated to second order by multiplicative terms, that is, product terms of two variables, the second term in a Taylor series expansion of the function.

2.5 MULTIPLICATIVE LATERAL INHIBITION: A DERIVATION

From the standpoint of a neural network and pre- or post-synaptic modulation of conductance, a nonlinear model possessing multiplicative properties can be developed. In Fig. 2.3 a circuit representing the electrical properties of an infinitesimal patch of nerve

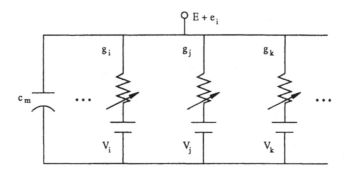

Figure 2.3   The circuit representing properties of an isopotential patch of neuron membrane, with variable conductances for ions of the given Nernst potentials, represented by the batteries. This circuit is one in a network of circuits interconnected via the conductances and described by Eq. (2.6) in the text.

membrane or an isopotential cell is shown. The membrane resting potential is E, on which there is a variation e_i arising from changes of the conductances g_i which are controlled by neighboring cell voltages. The batteries represent the Nernst potentials of the respective ion channels, and are the driving force for the circuit which is nonlinear because of the conductance inputs which in a network are also dependent on e. The membrane resting potential E is given by, for $g_j = \bar{g}_j$ a constant conductance,

$$E = \frac{\sum_j V_j \bar{g}_j}{\sum_j \bar{g}_j},\qquad (2.5)$$

and for voltage controlled conductance $g_j = \bar{g}_j + k_j e_j$ the set of network describing equations for the variations e on the resting potential E becomes:

$$C_m \frac{de_i}{dt} = -e_i \sum_j \bar{g}_j - e_i \sum_j k_j e_j + \sum_j k_j e_j (V_j - E), \ i = 1, 2, \dots. \qquad (2.6)$$

It is clear that there are linear cross-coupling inhibitory or excitatory terms which depend on the difference between the resting potential and the Nernst potential for the particular ion of that conductance. There are linear input terms given that an input $L_i(t)$ determines one of the conductances for each cell, in the form $g_i = \bar{g}_i + L_i(t)$:

$$\begin{aligned} C_m \frac{de_i}{dt} = &-e_i \sum_j \bar{g}_j - e_i \sum_{j \neq i} k_j e_j + \sum_j k_j e_j (V_j - E) \\ &+ L_i(t)(V_i - E) - e_i L_i(t), \ i = 1, 2, \dots \end{aligned} \qquad (2.7)$$

Of great importance is the fact that there are nonlinear cross-coupling terms of the form $k_j e_j e_i$, which are the multiplicative lateral inhibition, added to the linear lateral inhibition or excitation. A bilinear term for every input is $L_i(t) e_i$, and often in analysis this term has not been included on the basis that it is of the same order as the dominant nonlinearity, and an example of its effect can be seen in Fig. 2.4. Comparing the adaptive changes

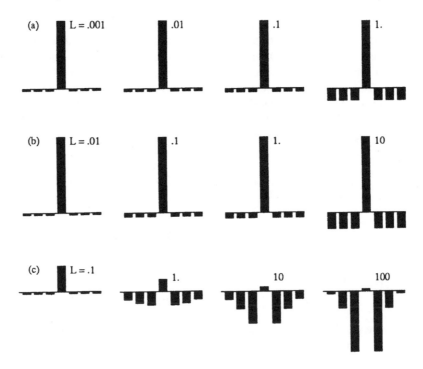

Figure 2.4 Three families of approximate receptive fields calculated from Eqs. (1.10) and (1.11), showing the excitatory center and inhibitory neighbor coefficients as a function of mean luminance L_0. In a, the result from Eq. (1.10) is shown, where as discussed in the text the input is a controlled conductance. In b, the feedforward term $L_i(t)e_i$ (in Eq. (1.7) is removed effectively making the input a controlled current source, otherwise the same as a. In c, the approximate receptive field is calculated for the nonuniform nonlinear lateral inhibition of Eq. (1.11). Scales: a, b coefficients normalized to excitatory center normalized to 1; c not normalized.

in Fig. 2.4a and Fig. 2.4b, there is not a strong difference shown by inclusion of the $L_i(t)e_i$ term.

When a variational analysis is performed, and linear terms retained, the multiplicative terms yield linear terms the coefficients of which depend on the mean luminance L_0, (see Eq. (2.11)). This form approximates the parametrically controlled second spatial differentiation invoked by Kelly (1975) to model the adaptive aspect of the human perceptual MTF. That differentiation is given by Eq. (2.4) above, with the assumed parameter $a(L_0)$ producing the adaptation by increasing the differentiation with increase in L_0. But in the nonlinear lateral inhibition model, the parametric dependence on L_0 is inherent in the system, distributed among the coupling terms, and not due to an attached parameter. The effect of the increased differentiation on the MTF is shown in Fig. 2.1. An equivalent set of MTF curves calculated from this model, Eq. (2.10), is shown in

Fig. 2.5, for a mean luminance L_0. For this calculation each input $L_i(t) = L_0 + l_i(t)$, and this model for the MTF consists of the spatial steady-state relationship given by Eq. (2.11), and a set of sinusoidal inputs $l_i(t)$ equivalent to a moving sinusoidal grating of the given spatial frequency. Using a Fourier transform on this set of equations, and assuming a fixed low pass characteristic for the physiological optics of the eye (Pinter 1985) the scaled MTFs for several mean luminances are shown in Fig. 2.5a. In this nonlinear sensory neural network model, the receptive field is spatio-temporal, and the steady-state spatial receptive field is given by the steady state relation for the variational e,

$$0 = -e_i \left(\sum_j \bar{g}_j + \sum_j k_j E_0 + L_0 \right) - \sum_j k_j e_j (E - V_j + E_0) + l_i(V_i - E - E_0), \quad (2.8)$$

or in matrix form as

$$0 = \mathbf{A}e + k\mathbf{l} \qquad (2.9)$$

This leads simply to a matrix inversion to obtain the receptive fields, the columns of the inverted matrix, $\mathbf{B}$.

$$e = -\frac{1}{k}\mathbf{A}^{-1}\mathbf{l} = \mathbf{B}\mathbf{l} \qquad (2.10)$$

However, a simpler approximation can also be utilized. Consider that the steady-state relation for a very large network can be written uniformly, for each cell i as

$$e_i = \frac{l_i(V_i - E - E_0)}{\left(\sum_j \bar{g}_j + \sum_j k_j E_0 + L_0 \right)} - \sum_j e_j k_j \frac{E - V_j + E_0}{\left(\sum_j \bar{g}_j + \sum_j k_j E_0 + L_0 \right)} \qquad (2.11)$$

Then the receptive field is in effect the set of coefficients of the excitation and inhibition on the right-hand side, and an insight into the nature of the mean luminance adaptation can be gained, without a matrix inversion, by examining the dependence of these coefficients on the mean level L_0. In Fig. 2.4a, these coefficients as receptive field are shown for the above coupling. For a gradient of nonlinear functions of coupling as:

$$K_j e_j \rightarrow f_j(e_j), \text{ and}$$
$$f_{\pm 1} = k_{11}e_{i\pm 1} + k_{12}e_{i\pm 1}^2 + k_{13}e_{i\pm 1}^3$$
$$f_{\pm 2} = k_{21}e_{i\pm 2} + k_{22}e_{i\pm 2}^2$$
$$f_{\pm 3} = k_{31}e_{i\pm 3}, \qquad (2.12)$$

the receptive field adaptation is shown in Fig. 2.4c. The main difference between these lies in the gradient of degree of nonlinear lateral inhibition for the latter case, which leads to a concentration of the inhibitory part of the receptive field toward the excitatory center, in Fig. 2.4c. This implies a stronger differentiation, less smoothing, and a movement of the MTF peak toward higher spatial frequencies, as the mean luminance increases.

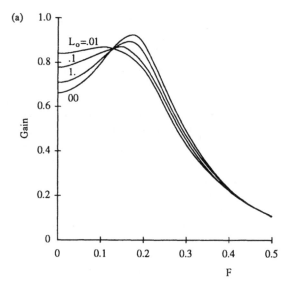

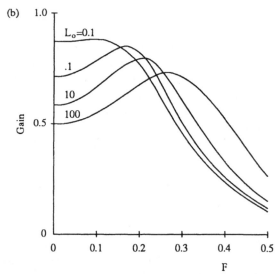

Figure 2.5 Two families of MTFs as a function of mean luminance L_0, where F is spatial frequency in cycles/degree of visual angle. In a, the MTF is calculated, by assuming sinusoidal spatial input in Eq. (2.10), for uniform nonlinear lateral inhibition to the third neighbor. In b, the nonuniform nonlinear lateral inhibition of 2.12 is used. In both cases a fixed low pass filter of the form $m = \exp(-9.11F^2)$ (Pinter 1985) is in cascade with the inhibitory network. Much sharper selectivity or relative peaking can be obtained by other choice of the fixed low pass filter, which is a function of the physiological optics and other factors in a visual system.

That is precisely what is shown in Fig. 2.5b, which has been calculated for the nonlinear lateral inhibitory coupling of Eq. (2.12).

For each mean luminance, the MTF of Fig. 2.5b and its receptive field in Fig. 2.4c is approximately the desired form of function that the application of linear prediction theory yields, for the purpose of estimating center excitatory values by integrating the inhibitory surround (Srinivasan, Laughlin, and Dubs 1982). The uniform lateral inhibition coupling giving the MTF of Fig. 2.5a and the receptive fields in Fig. 2.4a is a better approximation to the human perceptual data of Kelly (1975) (Fig. 2.1), however, since there the peak response frequencies do not shift with mean luminance, they remain constant.

In principle any finite receptive field profile, similar to library vectors in computational neural network theory, can be synthesized by the lateral inhibition equations using the general nonlinear coupling, in (2.12) the $f_j(e_j)$, and the mean luminance determines the convergence, or approximation to a given or desired receptive field. In the nervous system there are undoubtedly many types of receptive field profiles, optimized in some sense for a given task. For example, matched filtering in the classical sense (Siebert 1985) can be applied to spatio-temporal convolutions of known experimental records of impulse responses (the first order convolution kernels) of second-order (Lamina Monopolar Cells, LMC) visual interneurons in the fly retina to predict that the fly prefers edges at high luminances, and blobs or dark regions and not edges at low luminances (Pinter et al. 1990). The study of behavior of walking fly (*Lucilia*) confirms this prediction, with conclusions on which facets of the compound eye are mediating the behavior (Osorio et al. 1990). Further theory on these bases has been developed suggesting a very general approach to edge selectivity by the visual system (Srinivasan et al. 1990).

An interesting model of adaptive receptive field without inhibitory surround has been described by Cornsweet and Yellott (1985). The excitatory field adapts to higher luminance by narrowing and increasing its amplitude, and accounts well for many visual phenomena such as Mach bands and the change in MTF.

2.6 QUANTUM ABSORPTION EVENTS AND ADAPTIVE CHANGES IN LOW PASS FILTERING

Coarsening of sampling, or low-pass filtering, is inherent for visual systems operating at low photon absorption rates (shot-noise limiting process) (Barlow 1981). However, strategically self-adapting cellular mechanisms (such as nonlinear lateral inhibition) may also play a role for adaptation in the transition from low to high average luminance.

Photoreceptors' transmembrane potential response to the absorption of a photon is often termed the "quantum bump" (Fuortes and Yeandle 1964). The quantum bump changes its parameters depending on the mean rate of absorption of photons. Briefly, as mean luminance is raised, the bumps increase in mean rate, and become smaller in both amplitude and width, so that while at low luminances individual quantum bumps are visible, at high luminances they appear as a relatively small noise level on the steady level of generator potential of the photoreceptor (Wong and Knight 1980). The result of this is a somewhat underdamped photoreceptor temporal impulse response for adaptation at high luminance, but overdamped and slower or wider at low luminance. The photoreceptor impulse response and frequency response was demonstrated both directly and by Fourier transforms for *Limulus* retinular cells by Pinter (1966). In each case the photoreceptor is adapted to a given level of mean luminance. Thus, the temporal impulse response of photoreceptors adapts in a manner which supports or augments the spatial impulse response adaptation discussed above. As mean luminance increases, the spatio-temporal impulse response narrows, increasing bandwidth.

While the physical external quantum efficiency of a photoreceptor or visual interneuron may be greater than one on the basis of the number of charges gated through cell membranes for one photoisomerization of visual photopigment, the signalling of the photon absorption by one quantum bump or neural quantum usually implies an external quantum efficiency less than one. The example calculated by Barlow (1965) for nerve impulses (spikes) in a cat retinal ganglion cell shows a range of external quantum efficiency of 10^{-6} at high luminance up to 0.16 at low luminance. In this case the high luminance is higher than the low by a factor of one million (six decades or logarithmic units). In this case the signalling of quantal absorptions is embedded in a train of random phase spikes, so that noise due to photon absorptions is mixed with inherent transducer noise. This may be one factor in the observation that we do not perceive pointillism in low luminance images. Another factor may be a strategically self-adapting cellular mechanism such as nonlinear lateral inhibition in combination with inherent constant low-pass filtering. As the luminance increases, the differentiation property of lateral inhibition tends to compensate for the inherent low-pass filtering and increase the information capacity of the visual system. This occurs concomitantly with the increased information available at the peripheral level of the visual system due to the greater rate of photon absorptions at higher luminance levels.

3

Computational Properties

3.1 MODELS

Mathematical description of lateral inhibition makes it possible to relate and compare these network models with those previously investigated for their computational properties. The basic derivation was described in the previous chapter where it was shown that modulation of the conductance of a cell by other cells in a biological neural network can be depicted as in Fig. 2.3, redrawn in terms of variational parameters in Fig. 3.1, where x_i is the voltage variation of the ith cell, or an iso-potential portion of the cell, from its resting potential, C is the cell, or cell portion, membrane capacitance, R is the cell, or cell portion, membrane resistance, I_i are current inputs from sensors or antecedent cells, and $f_j(x_j)$ are conductances *controlled by neighboring cells' voltages*. Summing the currents at each node will result in a set of nonlinearly coupled differential equations

$$C\frac{dx_i}{dt} = I_i - \frac{x_i}{R} - x_i\Big(\sum_{j\neq i} f_j(x_j)\Big), \tag{3.1}$$

that can be written in the normalized form

$$\frac{dx_i}{dt} = I_i - A_i x_i - x_i\Big(\sum_{j\neq i} f_j(x_j)\Big), \tag{3.2}$$

which completely describe the dynamics of the system.

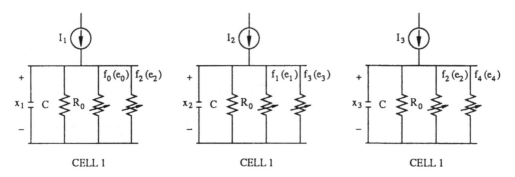

Figure 3.1 Multiplicative lateral inhibition via conductance modulation.

The crucial feature of this model is the control of conductances by neighboring cells which gives rise to the multiplicative terms $x_i f_j(x_j)$, and hence the name Multiplicative Lateral Inhibition (MLI). This separates the model from similar ones such as the model studied by Plonsey and Fleming (1969), which is of similar form but without the multiplicative term, or Baylor, Hodgkin, and Lamb (1974). In Curlander and Marmarelis (1983) a time-variant and space-dependent conductance establishes feed-back from horizontal cells to photoreceptors of vertebrate retinas, but is not directly controlled by neighboring cells. Interestingly, a similar model, without the input term and with positive $A_i x_i$, was proposed in population biology by Lotka (1956) over three decades ago.

In general, models in which multiplicative, or shunting,[1] terms of the form $x_i \sum_j I_j$, or $x_i \sum_j f_j(x_j)$ appear in the network equations are termed shunting neural networks, to be contrasted with *additive* models where excitatory or inhibitory interactions occur by addition or subtraction respectively. The first case shows non-recurrent, or feed forward, activity while the second is recurrent or with feedback. The polarity of these terms signify excitatory or inhibitory interactions.

In addition to this derivation which is based on passive membrane equation with synaptic interaction (Pinter 1983b), shunting nets have also been directly derived (Grossberg 1973) from the space-clamped version of the Hodgkin-Huxley equation (1952), or models of dendritic interaction (Rall 1977; Koch et al. 1982), or studies of directional selectivity (Torre and Poggio 1978) in the visual system. Ellias and Grossberg (1975) suggest that the experiments of Kuno and Miyahara (1969) on cat motoneurons is indicative of shunting interactions as well. Biological evidence of multiplicative interaction is detailed in Section 2.4.

Prior to describing some of the unique processing capabilities of these networks which make them of great computational interest, the closely related class of additive models will be briefly introduced.

---

[1] The terms shunting and multiplicative are used interchangeably here; a convention which is appropriate in view of the parallel paths of the circuit diagram of Fig. 3.1.

3.1.1 Additive Models

The models that are of the general form

$$\frac{dx_i}{dt} = -A_i x_i + \sum_j f_j(x_j) B_{ji} z_{ji}^{(+)} - \sum_j g_j(x_j) C_{ji} z_{ji}^{(-)} + I_i, \qquad (3.3)$$

can be termed *additive*, to be contrasted with multiplicative, or shunting, models. In Eq. (3.3) A_i assures exponential decay to resting level in the absence of any input to the network, f_j and g_j are signals from other cells in the network, B_{ji} and C_{ji} are connection strengths, and $z_{ji}^{(-)}, z_{ji}^{(+)}$ are Long Term Memory (LTM) traces which can change when the network *learns*. The first sum then formulates excitatory interactions while the second describes inhibitory interactions.

These additive models have historically been the pillars of neural network theory and have been shown to perform a variety of computational tasks as well as explain many biological, psychological, and psychophysical data. This is exemplified by the classic work of Hartline and Ratliff, which dates back to 1930s (Hartline et al. 1974; Ratliff 1965), on the lateral eye of *limulus*, which was described in Section 1.1. The seminal work of McCulloch and Pitts (1943), on the other hand, established a complete *logical calculus* for computation by *formal neurons* and thus provided a solid mathematical foundation for neural network theory. Rosenblatt (1958) used additive networks in multilayer learning architectures, the *perceptrons*, which founded the field of neural networks.

More recent works by Grossberg (1969, 1982, 1988b) explained a variety of psychological and psychophysical phenomena in terms of neural network equations and architectures, and explored the link between mathematics, biology and psychology. In two influential books Kohonen (1977, 1984) studied the associative memory properties of neural networks, and their application to classification problems, while keeping the biological feasibility of the models in sight. Anderson et al. (1977, 1983) used state vectors as elementary entities of cognitive computation and simulated a variety of psychological phenomena. Hopfield (1982, 1984) demonstrated and analyzed the global behavior of some networks by their ability to solve multi-constraint search problems. Ackley, Hinton, and Sejnowski (1985) exploited concepts from statistical mechanics to design a network capable of learning internal representations of various inputs.

Since the present work studies one-layer neural networks, multilayer learning architectures such as the back propagation networks (Rumelhart et al. 1986), or Adaptive Resonance Theory (ART) (Carpenter and Grossberg 1986), which attempt to model higher-order procesing such as learning are not treated here. Excellent survey articles (Levine 1983; Lippmann 1987; Grossberg 1988a) and collections of papers (Rumelhart and McClelland 1986; Anderson and Rosenfeld 1988; Koch and Segev 1989) can be consulted for thorough surveys of the field of neural networks and connectionist systems.

Furthermore, the classification of network models into additive and shunting, which follows Grossberg's convention (1988) is meant to highlight the distinctions by providing a comparative framework. The term *additive* should not, of course, be taken to mean *linear.* In these models, functions f_j and g_j are often either sharply nonlinear as in logical functions, or linear but thresholded, or sigmoidal. Indeed the shape of the nonlinearity specifies important characteristics of the networks. This is best seen in, for example, the "Boltzman machines" (Ackley et al. 1985) where a *temperature* parameter which determines the slope of the sigmoidal nonlinearity is varied via an *annealing schedule* and causes the network to learn an internal representation of the input.

Notwithstanding the rich heritage of additive models, shunting networks have important advantages over these models. In the next section some of these functional advantages are investigated.

3.2 PROPERTIES RELATED TO VISION

An impressive body of work addresses the general properties of shunting networks (Furman 1965; Grossberg 1982; Poggio and Torre 1978; Pinter 1983a, b). Specific properties applicable to the processing of visual information, especially low-level or peripheral processing, have also been extensively studied.

Pinter (1983a, b, 1984, 1985), as mentioned in the previous section, has used a network described by only shunting inhibitory connections, as in Eq. (3.1), to account for the short-term adaptive response of many visual units, such as the dependence of the receptive field organization on contrast and mean luminance, dependence of the spatial and temporal modulation transfer functions on contrast and mean luminance, dependence of size preference and latency of response time on contrast and mean luminance, and the dependence of incremental sensitivity on mean luminance.

Grossberg (1973, 1981, 1983, 1986b) has introduced shunting feed forward networks of the form

$$\frac{dx_i}{dt} = -A_i x_i + (B_i - x_i) \sum_k I_k C_{ki} - (x_i - D_i) \sum_k I_k E_{ki}, \qquad (3.4)$$

and equations for recurrent, or feedback, networks

$$\frac{dx_i}{dt} = -A_i x_i + (B_i - x_i) f_i(x_i) - x_i \sum_{j \neq i} f_j(x_j) + I_i, \qquad (3.5)$$

as well as network equations where excitatory and inhibitory inputs are separated and shunted

$$\frac{dx_i}{dt} = -A_i x_i + (B_i - x_i) \left[\sum_j f_j(x_j) C_{ji} + I_i \right] - (x_i - D_i) \left[\sum_j g_j(x_j) E_{ji} + J_i \right], \qquad (3.6)$$

where constants B_i and D_i specify the upper and lower bounds for the variation of activities x_i, A_i is the exponential decay term, and coefficients C_{ji} and E_{ji} describe the fall-off with the distance between cells x_k, and x_i of the excitatory and inhibitory influences, respectively, of input I_k on cell x_i.

These networks are then shown to explain a host of visual phenomena such as the Cornsweet and Craik-O'Brian effects, phantoms and subjective contours, binocular brightness summation, multiple spatial frequency scaling and edge detection, figure-ground completion, coexistence of depth and binocular rivalry, reflectance rivalry, Fechner's paradox, decrease of threshold contrast with increased number of cycles in a grating pattern, adaptation level tuning, Weber law modulation, and the shift of sensitivity with background luminance.

Cohen and Grossberg (1984), Grossberg and Mingola (1985, 1987) and Grossberg and Todorovic (1988), have used equations (3.4–3.6) in conjunction with other mechanisms to explain a wide variety of higher level brightness perception phenomena such as perceptual segmentation and grouping. Such phenomena are used in the investigation of how a continuous image is filled-in, or completed, after being coded by discrete components of a neural network. These workers also suggest areas of the visual cortex where such processes may occur and contend that no other alternative visual theory has been capable of explaining as wide a range of properties as the shunting networks. A general-purpose model of preattentive vision which includes shunting nets as a primary module of the architecture is described by Grossberg et al. (1989).

Specific processing capabilities of these networks which are especially suitable for technological applications, such as edge enhancement, feature extraction, data compression, adaptation to mean luminance, tunability of sensitivity, capability to code the intensity, content addressable memory property, high-order classification, and, in short, network's capability to establish a *good representation* of the input data have been reported in Nabet and Darling (1988), Nabet et al. (1989, 1990) Darling and Nabet (1988), and Darling et al. (1989), Nabet (1989, 1990) and will be further discussed here.

In view of the very broad applicability of shunting networks and their many interesting properties, the fact that there exists a framework for very simple implementation of these networks, as will be demonstrated in the present work is of special significance.

In the following subsections some of the processing capabilities unique to networks with multiplicative inhibition will be reviewed. The emphasis will be on intuition rather than rigorous mathematics and the discussion is based on references mentioned above.

3.2.1 Variable "Connection Strengths"

Consider the simple additive model

$$\dot{x}_i = I_i - \frac{x_i}{R} + \sum_j T_{ij} x_j, \tag{3.7}$$

where T_{ij}'s are connection strengths, which gained fame in a paper by Hopfield (1984) marking the resurgence of interest in neural networks. A shunting version of this model can be written as

$$\dot{x}_i = I_i - \frac{x_i}{R} + x_i \sum_j T_{ij} x_j, \tag{3.8}$$

or in the form

$$\dot{x}_i = I_i - \frac{x_i}{R} + \sum_j T_{ij}(x_i) x_j. \tag{3.9}$$

That is, the connection strengths are now variable (state-dependent). Given that these connection strengths specify the global dynamics of the system, for example the locations of the memories in an associative memory system, their variability increases the complexity of the dynamical behavior of the system. This raises the following questions:

- Does this increase of complexity translate into an increase in processing capabilities of the network?

- Do the computational and implementational costs of having state-dependent connection strengths justify the advantages achieved?

The discussion of the previous section provides a positive answer to the first question and in Chapters 5 and 6 it will be shown that with correct circuit design techniques a multiplication can be achieved by only *one* transistor and thus the circuit which implements a shunting network can be simpler than a similar additive model.

3.2.2 Automatic Gain Control

In the steady state, the existence of which will be shown in Chapter 8, the MLI Eq. (3.2) can be written as

$$x_i = \frac{I_i}{a_i + \sum_j f_j(x_j)}. \tag{3.10}$$

The stability, automatic gain control, and wide dynamic range properties of such feedback systems are well known in control theory.[2] These properties make the network very suitable for stable short term memory storage. Such a gain control mechanism in conjunction with carefully chosen nonlinearity in the feedback loop, that is, the shape of $f_j(x_j)$, makes the network sensitive to small input values by suppressing noise while not saturating at high input values; it thus solves what Grossberg (1973) has termed *the noise–saturation, or stability–plasticity, dilemma*. Additive models lack this simple gain control mechanism and can not solve the noise–saturation dilemma.

3.2.3 Some Psychophysical Properties

Consider the input pattern

$$I_i = \theta_i I_t, \tag{3.11}$$

where I_t is the total input, defined by

$$I_t = \sum_{k=1}^{n} I_k. \tag{3.12}$$

The θ_i then specify normalized *contrast* ('normalized' since from Eq. (3.11), $\sum_{i=1}^{n} \theta_i = 1$) or *reflectances*, while I_t parameterizes *total activity, background or ambient intensity* or, if divided by the number of cells, *mean luminance* levels.

_____

[2]See for example D'Azzo and Houpis (1978).

Now, in the original derivation of Eq. (3.1), Pinter (1983b) notes that the input can actually be provided by a variable conductance. Using a first approximation, variable conductances can be replaced by input current sources to get the steady-state form of Eq. (3.1)

$$x_i = \theta_i \frac{I_t}{a + \sum_{k \neq i} I_k} \tag{3.13}$$

Further assuming that the total activity is not greatly perturbed by the activity of one cell; that is,

$$\sum_{k \neq i}^{n} I_k \approx \sum_{k=1}^{n} I_k = I_t, \tag{3.14}$$

Eq. (3.13) can be written as

$$x_i \approx \theta_i \frac{I_t}{a + I_t}. \tag{3.15}$$

Equation (3.15) is a form of the Weber-Fechner law of visual psychophysics, which shows how the reflectances are modulated by mean intensity. This helps explain aspects of an interpretation termed *brightness constancy* (Cornsweet 1970).

On the other hand, the total activity of the network is

$$X_t = \sum_{i=1}^{n} x_i = \sum_{i=1}^{n} \theta_i \frac{I_t}{a + I_t} = \frac{I_t}{a + I_t}, \tag{3.16}$$

which reaches 1 as mean intensity (background) increases. This serves to show that the total activity of the network is automatically normalized even as the input grows without bound. Also, as a result, if a part of the network is very active (bright) this has to be at the cost of another part becoming inactive (dark). In visual psychophysics this has been called *brightness contrast* (Cornsweet 1970). Grossberg (1983) notes that it is very interesting that the two seemingly contradictory phenomena of brightness constancy and brightness contrast can be explained by the same network.

Finally, the modifications that led to Eq. (3.15) have transformed the network Eq. (3.2) into the shunting feed forward equation

$$\dot{x}_i = -A_i x_i + (1 - x_i)I_i - x_i \sum I_k, \tag{3.17}$$

which is a subset of Eq. (3.4), some of whose properties, including the above, were discussed in the previous section.

3.2.4 The Coding of Intensity

The importance of the interaction of intensity and contrast, that is the dynamics of *mass-action*, is obvious: a light is turned on to read a book even though the contrast of the pages is independent of total intensity. The nervous system uses the larger number of photons available to better process the image. This is seen in Weber law Eq. (3.15) which shows how the reflectances are modulated by the intensity, and can also be observed if Eq. (3.10) is rewritten explicitly in terms of intensity and reflectance as

$$x_i = \frac{\theta_i I_t}{a_i + \sum_j f_j(x_j)}. \tag{3.18}$$

These equations also demonstrate how the intensity itself has been coded. Each cell has a "clue" to the total intensity without having to saturate its dynamic range by coding the intensity directly. This property is a direct result of dividing the intensity by either $(a_i + I_t)$ in Eq. (3.15) or by $(a_i + \sum_j f_j(x_j))$ as in Eq. (3.18), and is due to the unique automatic gain Control property of shunting nets. Front-end processors which lack this property should either operate in very controlled lighting environments or normalize the input; either case can destroy an important feature of the data.

4

Nonlinear Lateral Inhibition and Image Processing

M. E. Jernigan, R. J. Belshaw, G. F. McLean

4.1 INTRODUCTION

Models of lateral inhibition in biological vision systems have long been recognized as sharing both form and function with operations of digital image processing. Barlow describes the function of lateral inhibition as to suppress response to "usual" signals, to improve the detectability of useful signals, and to protect the informative signal from degradation during transmission and manipulation in the vision system (Barlow 1962; Sakitt and Barlow 1982). The Mach band phenomenon in psychophysics is a manifestation of the edge enhancement capabilities of the simplest linear models of lateral inhibition. Hubel and Wiesel's feature detectors have receptive fields which exhibit the familiar antagonistic center and surround of lateral inhibition (Hubel and Wiesel 1962). Digital image processing has the same goals: encoding, enhancement and feature extraction.

Image coding exploits local redundancy or correlation among picture elements to achieve data compression for transmission and storage. Srinivasan has noted that the adaptation of receptive fields in lateral inhibition resembles the instantaneous impulse response of an all-zero linear predictive coding mechanism used for video bandwidth compression (Srinivasan et al. 1982). On the other hand, the global dynamic range reduction associated with the logarithmic sensitivity to local intensity differences is not achievable with linear models of lateral inhibition. Digital image processors have evolved homomorphic systems with an explicit log transformation that matches the intrinsic multiplicative structure of images (Stockham 1972).

Image enhancement uses linear high frequency emphasis spatial filters whose impulse responses are identical in form to the lateral inhibition response profile and whose frequency responses are narrow band spatial frequency channels. Wilson and Bergen describe these mechanisms as difference of gaussians in their four channel model (Wilson and Bergen 1979). Marr's efforts to bridge natural and machine vision with their common fundamental problem led to a gaussian smoothed laplacian operator with similar spatial and frequency domain characteristics as a first stage in edge detection (Marr et al. 1980). More recent descriptions of these mechanisms invoke Gabor filters (gaussian modulated sinusoids in space, gaussian band pass filters in frequency) which show additional positive and negative surrounds beyond the initial inhibitory region. Regardless of the particular form, these linear enhancement filters are limited in effectiveness by their inability to adapt to local image characteristics: varying illumination and varying local signal to noise ratio. Modern image processing resorts to a family of more or less ad hoc adaptive and nonlinear enhancement filters that are tailored to specific degradations.

Finally, feature extraction is typically either edge based, using edge detectors or region based, using local texture properties for segmentation. Marr's edge detection scheme applies a zero crossing detector to the gaussian smoothed laplacian and is dependent on heuristically determined thresholds and multiple channel coincidence for reliable edge detection. Optimal edge detectors in digital image processing have responses that are asymptotically identical to both Wilson and Bergen's difference of gaussians and Marr's operator (Jernigan et al. 1981; Shanmugam et al. 1979). Texture feature extraction can also be viewed as a local spatial frequency filtering using Gabor functions to extract frequency, orientation, and phase of local texture components (Wright and Jernigan 1986).

As the limitations of the linear approaches to image coding, enhancement, and feature extraction became apparent, research in image processing began to disperse into the three goal-driven directions. A variety of distinct adaptive and nonlinear schemes emerged in each area in an effort to address the underlying nonadditive, nonstationary character of images. Modern enhancement methods simultaneously compress dynamic range and sharpen edges. Edge-adaptive smoothing allows for noise filtering while preserving signal detail.

Nonlinear lateral interaction in the retina leads to image processing which adapts to local image characteristics in ways that are qualitatively similar to successful adaptive filtering strategies in digital image processing (Peli and Lim 1982). Specifically, the adaptation to local luminance described by Pinter in his multiplicative lateral inhibition model shows an increased edge enhancement as local mean intensity increases, as well as an overall gain adaptation (Pinter 1983). More generally, Pinter's nonlinear lateral inhibition model simultaneously addresses problems of coding, enhancement, and extraction, as it acts to compress the dynamic range, reorganize the signal to improve visibility, suppress noise, and identify local features. More specifically, the following characteristics emerge:

- There is a local gain adaptation, which serves to compress the dynamic range of the signal and enhances *local* variations when there is a wide *global* dynamic range.

- There is a local shape adaptation, which can be viewed as a detail preserving adaptive filter. This filter sharpens when the local signal characteristics are larger than the assumed additive noise characteristics.

- There is a relative edge enhancement, once again linked to the certainty of a local feature being some structural element as opposed to some association of local noise.

Finally, Pinter's model is a one-step process, (i.e., it does not require an estimate of local properties, followed by filter tuning as is common with current adaptive filters), it is applied in parallel with no a priori knowledge of the image, it requires no forced choice, and it satisfies all three functions of image processing. Although the solution of the model is difficult and time-consuming on general purpose computers, we find its capabilities intriguing from the perspective of the development of truly general machine processes for the manipulation of images. Our purpose in this forum is to cast Pinter's model in a discrete signal processing framework and examine its behavior as a digital image processing system.

4.2 BACKGROUND

In Pinter's formulation of nonlinear lateral inhibition, the time derivative of the output of the nth cell is related to the input of the nth cell and a product of the output with a function of the outputs of neighboring cells:

$$C\frac{dy(n)}{dt} = x(n) - y(n)\left[1 + \sum_{k \neq n} f_k\left(y(k)\right)\right] \qquad (4.1)$$

where $x(n) =$ input of the nth cell
$\quad\quad\quad\quad y(n) =$ output of the nth cell
$\quad\quad\quad\quad\quad C =$ a constant (membrane capacitance)
$\quad\quad f_k\left(y(k)\right) =$ a function of the output of the neighbouring cell at position k.

In one version, these functions of neighbouring outputs were polynomials whose order decreased as the distance to the neighbour increased. For example,

$$f_{n\pm 1}\left(y(n\pm 1)\right) = a_{11}y(n\pm 1) + a_{12}y^2(n\pm 1) + a_{13}y^3(n\pm 1)$$

$$f_{n\pm 2}\left(y(n\pm 2)\right) = a_{21}y(n\pm 2) + a_{22}y^2(n\pm 2)$$

$$f_{n\pm 3}\left(y(n\pm 3)\right) = a_{31}y(n\pm 3).$$

The nonlinear and recurrent nature of the model lead to adaptive edge gain as well as reduced gain in constant intensity regions as described in Pinter (1984). Preliminary experiments of a two dimensional implementation yielded results in qualitative agreement with the one dimensional results and suggested the model's image enhancement potential (McLean et al. 1988). In an effort to better understand the essential structure and behavior of the model we have been studying alternative and simplified implementations.

To study the steady state behavior, set $\frac{dy(n)}{dt} = 0$, and the model becomes,

$$y(n) = \frac{x(n)}{1 + \sum_{k \neq n} f_k(y(k))}. \tag{4.2}$$

The polynomials of decreasing order may perhaps be replaced by an exponential form,

$$f_k(y(k)) = a_k \left[e^{y(k)/d_k} - 1 \right] \tag{4.3}$$

where a_k = constant for the kth neighbor, and d_k = distance measure to the kth neighbor. The similarity to the polynomial form is apparent with the series expansion,

$$e^{y(k)/d_k} = 1 + \frac{y(k)}{d_k} + \frac{y^2(k)}{2d_k^2} + \frac{y^3(k)}{6d_k^3} + \cdots,$$

where nearer neighbors (smaller d_k) effectively include higher powers of $y(k)$.
As a nonlinear, discrete system,

$$y(n) + y(n) \left[\sum_{k \neq n} f_k(y(k)) \right] = x(n).$$

If the effect of all but the nearest neighbor is ignored and the exponential is approximated by the linear term, the model simplifies to

$$y(n) + ay(n)\left[y(n-1) + y(n+1)\right] = x(n), \tag{4.4}$$

where c represents the various constants and neighbor effects are assumed to be symmetric. For constant or DC inputs the response is constant and $y(n) = y(n+1) = y(n-1)$. The equation can thus be reorganized into a familiar quadratic equation. In that case $2ay^2(n) + y(n) - x(n) = 0$ and

$$y(n) = \frac{\sqrt{1 + 8ax(n)} - 1}{4a} \tag{4.5}$$

(constrain $y(n) > 0$ for $x(n) > 0$). To obtain a real valued output,

$$a > -\frac{1}{8x(n)}. \tag{4.6}$$

When applied to inputs containing edges, the system smoothes edges when $a < 0$ and sharpens edges when $a > 0$.

While greatly simplified, such a model retains both nonlinear and recurrent properties. In this form it can be compared to a linear second order difference equation whose behavior is well known, that is,

$$y(n) + by(n-1) + by(n+1) = x(n).$$

This system has an exponential impulse response of the form

$$h(n) = kc^{|n|}$$

where

$$c = \frac{-1 + \sqrt{1 - 4b^2}}{2b}.$$

It acts as a high pass filter for $-1 < c < 0$ ($0 < b < \frac{1}{2}$) and as a low pass filter for $0 < c < 1$ ($-\frac{1}{2} < b < 0$), and is stable for $|b| < \frac{1}{2}$.

In the remainder of this chapter we describe an implementation of this stripped down Pinter model, in one and two dimensions, discuss its behavior as an image processor, and compare it to more complete implementations.

4.3 IMPLEMENTATION

The more complete Pinter model using the exponential functions of Eq. (4.3) and a simplified version of the Pinter model following the direction of Eq. (4.4) were implemented.

The simplified one dimensional Pinter model was implemented by recursively calculating the following equation:

$$y_{k+1}(n) = \frac{x(n)}{1 + ay_k(n-1) + ay_k(n+1)} \tag{4.7}$$

where $y_0(n) = 1.0$ for all n works adequately.

When implementing this equation it was noted (for values of the parameter $0 < a < 1$) that iteration solutions tended to oscillate about the actual final solution or oscillate about a path on their way to a final solution. A simple modification of averaging two consecutive estimates greatly dampened the oscillation and reduced the total computation to final solution by at least 30 per cent.

The new iteration estimate z_m is calculated as follows:

$$y_{2m} = \frac{x(n)}{1 + az_m(n-1) + az_m(n+1)} \tag{4.8}$$

$$y_{2m+1} = \frac{x(n)}{1 + ay_{2m}(n-1) + ay_{2m}(n+1)} \tag{4.9}$$

$$z_{m+1} = (y_{2m} + y_{2m+1})/2 \tag{4.10}$$

where $z_0(n) = \ln(x(n))$ for all n.

The stripped down two dimensional Pinter model implementation involved using a 3×3 mask operator:

$$\begin{matrix} a_{11} & a_{12} & a_{13} \\ a_{21} & a_{22} & a_{23} \\ a_{31} & a_{32} & a_{33}. \end{matrix}$$

During our experiments the mask takes the form,

$$
\begin{array}{ccc}
0.0 & 0.3 & 0.0 \\
0.3 & 0.0 & 0.3 \\
0.0 & 0.3 & 0.0
\end{array}
$$

so that only the very nearest neighbors in both the row and column directions (x and y) have any effect on the equation solution. This corresponds with the development from Eq. (4.9) in which the parameter 'a' takes on the specific value 0.3 for our experiments.

This mask was effectively moved about the image, centered over each pixel location (in exactly the same way that a convolution mask would be), and multiplied with the pixel values which held corresponding locations with respect to the current center pixel. The sum of these products corresponds to the sum in Eq. (4.2), and is,

$$
\sum_{k \neq n} f_k(y(k)) = \sum_{i=1}^{i=3} \sum_{j=1}^{j=3} a_{ij} y(\text{row} + i - 2, \text{col} + j - 2).
$$

The steady state output solution for an image pixel value at the current row and column is represented as:

$$
y(\text{row}, \text{col}) = \frac{x(\text{row}, \text{col})}{1 + \sum_{i=1}^{i=3} \sum_{j=1}^{j=3} a_{ij} y(\text{row} + i - 2, \text{col} + j - 2)}
$$

This formula then can be used iteratively to solve for the steady state solution of the stripped down Pinter model.

The iteration estimate z_m was calculated in an analogous manner to that for the 1D case:

$$
y_{2m} = \frac{x(\text{row}, \text{col})}{1 + \sum_{i,j} a_{ij} z_m(\text{row} + i - 2, \text{col} + j - 2)} \tag{4.11}
$$

$$
y_{2m+1} = \frac{x(\text{row}, \text{col})}{1 + \sum_{i,j} a_{ij} y_{2m}(\text{row} + i - 2, \text{col} + j - 2)} \tag{4.12}
$$

$$
z_{m+1} = (y_{2m} + y_{2m+1})/2 \tag{4.13}
$$

where $z_0(\text{row}, \text{col}) = 1.0$ over all the rows and columns works effectively.

In order to check for convergence the following calculation was made:

$$
\rho_m = 100 \max_{\text{row,col}} \left[\frac{z_m(\text{row}, \text{col}) - z_{m-1}(\text{row}, \text{col})}{z_m(\text{row}, \text{col})} \right]
$$

which calculates the maximum percentage change of the previous estimate from the new estimate with respect to this new estimate. The iterations are considered to have converged to a solution when the maximum percentage change is less than 0.5 per cent ($\rho < 0.5$).

The more complete Pinter implementation involved summing functions of the form in Eq. (4.3) (all $a_k = 1.0$), where the sum is:

$$\Omega(y(r,c)) = \sum_{i=1,i\neq4}^{i=7} \sum_{j=1,j\neq4}^{j=7} \left[e^{(y(r+i-4,c+j-4)/d(i,j))} - 1 \right], \tag{4.14}$$

$$d(i,j) = 255\sqrt{(4-i)^2 + (4-j)^2}, \tag{4.15}$$

where the value 255 ensures that the exponent does not exceed the value one for input pixel values in the range 0 to 255. The solution can be calculated as:

$$y_{2m} = \frac{x(\text{row}, \text{col})}{1 + \Omega(z_m(\text{row}, \text{col}))}, \tag{4.16}$$

$$y_{2m+1} = \frac{x(\text{row}, \text{col})}{1 + \Omega(y_{2m}(\text{row}, \text{col}))}, \tag{4.17}$$

$$z_{m+1} = (y_{2m} + y_{2m+1})/2, \tag{4.18}$$

where $z_0(\text{row}, \text{col}) = \ln(x(n))$ over all the rows and columns and the convergence criterion remains identical to that for the simpler Pinter model. In fact the equation $z_0(\text{row}, \text{col}) = p + \ln(x(\text{row}, \text{col})) * [q + r * \ln(x(\text{row}, \text{col}))]$ models the DC normalization curve very well as a first approximation (p, q, r to be determined).

The execution time of the algorithm is also greatly diminished if the initial image estimate is created by using the DC-in to DC-out array (Fig. 4.4—previously calculated) to assign the corresponding DC-out values to the pixels. This preliminary processing speeds up the convergence as most real world images contain many regions of constant intensity. Ultimately the processing scheme may be implemented with parallel machines.

The more complete Pinter implementation took roughly nine times as long to converge to a solution as the simpler Pinter model.

4.4 RESULTS AND DISCUSSION

Unlike simple models of lateral inhibition which cannot explain all three facets of Barlow's functional model, Pinter's model of lateral inhibition using nonlinear local interactions produces a richness of response which suggests the simultaneous operation of all three aspects. In this section solutions of the 2D model are used to illustrate the potential for the Pinter model to be used for all three facets of image processing.

We are interested in the manner in which the model reorganizes the signal to improve visibility, suppress noise, compress the dynamic range and identify visual features. In Pinter's work, the model has been analyzed by performing a residual analysis of some small perturbation from a steady state stimulus (Pinter 1984, 1985). Here, we are interested in evaluating the overall response of the recursive, nonlinear model.

Figures 4.1 and 4.2 show the relative gain adaptation of the exponential and simplified Pinter models respectively. These graphs were obtained by measuring the DC

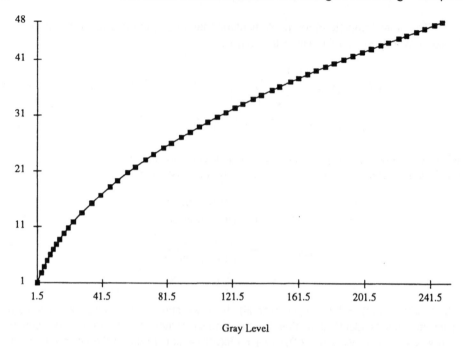

Gray Level

Figure 4.1   Exponential Pinter model DC value out vs. mean DC value in.

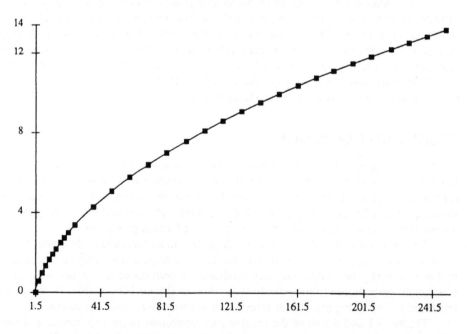

Figure 4.2   Simplified Pinter model DC value out vs. mean DC value in.

response of the models over a range of input values. Note that the responses are logarithmic, suggesting exactly the sort of dynamic range compression observed in natural vision systems, as well as the type of signal transformation associated with homomorphic processing.

Normalized response of the models to a unit impulse at various base gain values as the local mean increases leads to a relative sharpening of the filter response and relatively more ringing. Assuming an additive noise model (as is the case for neural noise in biological vision, Srinivasan et al. 1982), then lower mean values imply a lower signal to noise ratio. At higher means, the sharpness is increased and the gain also, so as to produce relative edge enhancement in the vicinity of strong, unambiguous edge features.

For image enhancement applications it is sometimes useful to normalize transformations to preserve the absolute intensities of relatively constant (DC) regions while enhancing the surrounding edges. The nonlinear transformation of mean DC value in to DC value out (see Fig. 4.1—Full Pinter model, and Fig. 4.2—Simplified Pinter model) can be used to output DC values at their original input value if desired. The lower DC values are transformed in a strong nonlinear manner while the equations converge quickly at this DC level (approximately 6 iterations). The higher DC values (greater than 50) are transformed in an almost linear manner while the equations take more than twice as long to converge (greater than 10 iterations).

Relative Edge Enhancement (REE) is defined as:

$$REE = \frac{Peak.value - Valley.value}{DC.high.background - DC.low.background},$$

and operates as a measure of the actual edge enhancement to a particular step size at a particular DC mean value. See Figs. 4.3 and 4.4 for plots of the relative edge enhancement versus mean DC value input for the more complex and simpler Pinter implementations respectively. Note specifically that the relative edge enhancement for differing step sizes becomes independent of step size near a DC value of 55 and takes on a more linear aspect thereafter (for both the simplified and more complex Pinter model). Note that of the three curves in each graph the superior curve is achieved with an edge step size of value one, while the inferior curve is achieved with an edge step size of 21.

Figures 4.5 and 4.6 respectively show the output to an image step input of the more complex and simpler Pinter implementations. The test image is a 32×32 image with a lower DC region of value 50 (background that is darker) and a higher DC region of value 150 (foreground that is lighter). These figures allow an appreciation for the region affected by the response to the step edge. Note in particular the peak and valley on either side of the edge, as well as the slight dip near the peak and the slight hump near the valley. Note that the simpler Pinter model affects significantly only the two nearest neighbors to the edge as would be expected with a nearest neighbor 3×3 mask, while the more complex Pinter model affects the four nearest neighbors to the edge as would be expected with a 7×7 mask that decreases exponentially away from the center. These figures illustrate the overall response of the models to a typical test signal. The gain adaptation and relative edge enhancement are both evident in the output of the models.

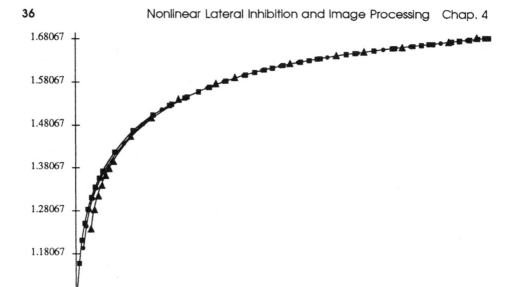

Figure 4.3   Exponential Pinter model Relative edge enhancement versus mean DC value in.

It is tempting to consider using the output of the Pinter model as the input to some form of edge detection. This scheme would show an improved normalization of response and decreased sensitivity to noise (due to the relative edge sharpening).

Figures 4.1 and 4.2 demonstrate how the Pinter model has compressed the entire signal (DC logarithmic transformation). Figures 4.5 and 4.6, while providing evidence of the compression, also demonstrate the important edge enhancement effect. This effect improves the detectability of the useful signal (the informative edge) while simultaneously suppressing the "usual" signal (DC background).

A checkerboard 32×32 test image with 7×7 checks was also processed by both models. Figures 4.7 and 4.8 respectively show the outputs from the more complex and the simplified Pinter model. The test image had lower DC checks of value 50 (darker background) and higher value DC checks of 150 (lighter foreground). Note the greater response of the model near the check corners, especially that of the simplified model. Note that the exponential Pinter model (using a 7×7 mask) affects almost each entire 7×7 check as would be expected, while the 3×3 mask simpler model has approximately a 5×5 DC response region in the interior of each check.

Figure 4.9 is a real world input image (192×200; 128 grey levels showable in range 0 to 255; here: 0 to 196), and Fig. 4.10 shows that same image after it has been processed by the raw exponential Pinter model (output in range 0 to 50).

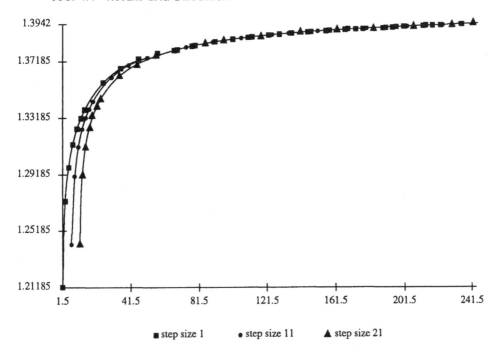

Figure 4.4 Simplified Pinter model Relative edge enhancement versus mean DC value in.

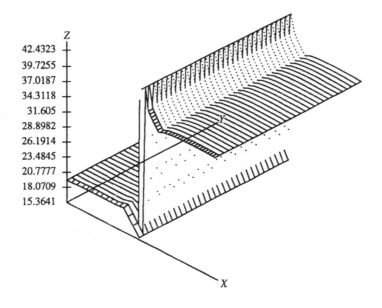

Figure 4.5 Exponential Pinter model Step edge response.

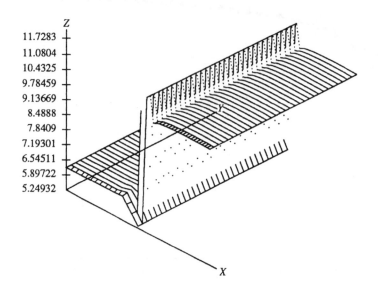

Figure 4.6　Simplified Pinter model Step edge response.

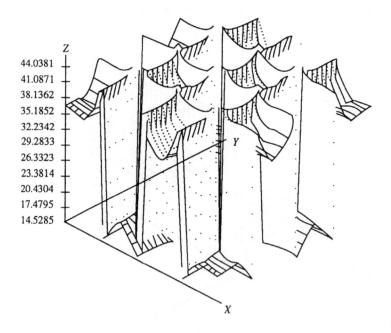

Figure 4.7　Exponential Pinter model Response to checkerboard.

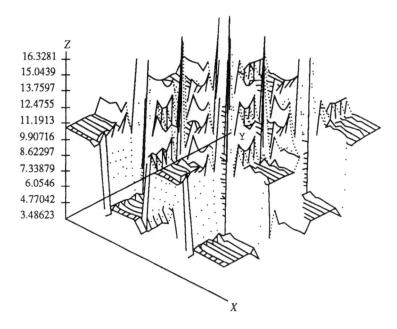

Figure 4.8   Simplified Pinter model Response to checkerboard.

Figure 4.9   Original image.

Figure 4.10   Exponential Pinter model Response to image. (Unscaled)

A reasonable approach to achieving edge enhancement while maintaining regions of constant intensity involves scaling the input image into some DC range where the DC normalization curve is relatively linear and where the relative edge enhancement is practically independent of edge step size and as constant as possible over the region. Noting also that the equations take longer to converge at higher DC values, it is prudent to choose as low a DC level as possible for the range. The range of values from 60 to 80 is an acceptable range for processing the image to achieve constant relative edge enhancement irrespective of edge strength, close to linear DC response and a characteristic stable Mach band type edge response throughout the range. Details of this normalization scheme may be found in (Jernigan et al. 1989).

Figures 4.11 and 4.12 are the output images from such a processing scheme utilizing the exponential and the simplified Pinter model respectively (clipped for display into the range 0 to 255). Note the improved facial features, the cleaner detail to the hair, the clarity of the wood grain in the chair, and the sharper shirt stripes and folds in the enhanced images. The fine texture of Fig. 4.12 has been affected by the simplified Pinter model. The texture contains patches of superimposed low amplitude cross-hatching or checkerboard patterns which themselves are grouped into a checkerboard pattern on a larger scale (Fig. 4.8 provides evidence that the non-symmetric region of support in this model leads to an exaggerated corner response).

Figure 4.13 shows a step edge response of the more complex Pinter model using the DC normalization algorithm. Note that the DC values correspond closely to their original DC input values (50—background, 150—foreground).

Figures 4.14–22 show the capability of the Pinter model in various modes. Figure 4.14 shows the original image with the road and desert in the foreground, the mist-

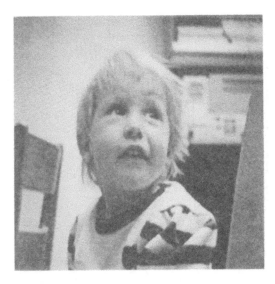

Figure 4.11 Exponential Pinter model DC value out to DC in preservation.

Figure 4.12 Simplified Pinter model DC value out to DC in preservation.

covered mountains in the background (original image range 0–195). Figure 4.15 is the same image scaled into the 0–255 range. Figure 4.16 is the original image processed by the exponential Pinter model and displayed in raw form (0–48). Figure 4.17 is Fig. 4.16 scaled up into the full 0–255 range. Figure 4.18 is the Pinter processed DC normalized

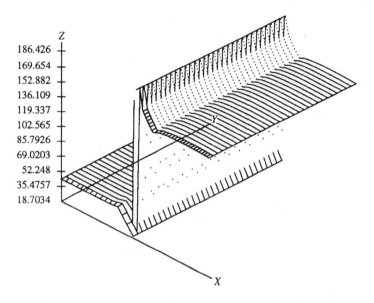

Figure 4.13 Exponential Pinter model Step edge response with DC value out to DC in preservation.

Figure 4.14   Original image.

version of Fig. 4.16 (exponential model; −24 to 210). Figure 4.19 is Fig. 4.18 scaled up into the full 0–255 range.

Figure 4.20 is the original image minus the laplacian of the original image (operator: one minus laplacian; −83 to 255). Figure 4.21 is Fig. 4.20 scaled up into the full 0–255 range. Figure 4.22 is the laplacian of the original image (range of −152 to 121; scaled into the 0 to 255 range for viewing).

Our image display program clips image gray values less than zero by replacing them with a value of zero (black). The clipping of the images presented is insignificant and does not affect our comparison (i.e., the number of pixels less than zero in the unscaled images is a negligible proportion of the total number of image pixels and this clipping occurs in near zero (black) regions of the image).

Figure 4.15   Original image scaled into the 0 to 255 range.

Figure 4.16   Exponential Pinter model.

Figure 4.17   Exponential Pinter model scaled into the 0 to 255 range.

Considering the limitations upon image comparison it should be noted that the scaling algorithm is based upon knowledge of the maximum and minimum gray values in the original unscaled image (which naturally vary from image to image). This fact implies that comparison is less obvious across scaled images (due to confounding scaling misalignments) and that comparison can more clearly be carried out across the unscaled images.

Note that Fig. 4.17 has a contrast-enhanced foreground (more detail visible), while the background contrast has been reduced and the edges have been simultaneously sharpened. This observation makes sense in light of the previously described curves. The lighter regions (high DC values) will be edge sharpened to a greater extent while the contrast will be less enhanced due to the decreasing slope of the logarithmic DC normalization curve.

Figure 4.18   DC value out to DC in preservation.

Figure 4.19   Exponential Pinter model DC value out to DC in preservation scaled up into the 0 to 255 range.

Figure 4.20   One minus laplacian of the original image.

The outline of the misty mountains in Fig. 4.18 is much more visible than in the original due to edge sharpening and the maintenance of the original contrast. The foreground brush detail and structure in Fig. 4.18 is much clearer than in the original image.

Figure 4.20 is the traditionally enhanced image (unscaled). Note that the mountain outline is sharper than in either the DC normalized image (Fig. 4.18) or the original (Fig. 4.14). The foreground of Fig. 4.20 is sharper than the original though the character of the original might be viewed to have been lost (superimposed high frequency structure of too high an amplitude resulting in distortion). The nonlinear nature of the Pinter model promises subtle adaptive response behavior rather than a crude linear response.

In this chapter we have examined a nonlinear lateral interaction which adapts to local image characteristics. The models studied show an increased edge enhancement as the local mean increases and also show a gain adaptation (logarithmic transformation).

Figure 4.21   One minus laplacian of the original image scaled up into the full
0 to 255 range.

Figure 4.22   Laplacian of the original image scaled into the 0 to 255 range.

The models have been formulated in a discrete signal processing framework and their
behavior as digital image processing systems has been evaluated. The simplified model
demonstrates the major characteristics of interest for this class of model and allows for
efficient exploration and experimentation of these characteristics as the model parameters
change. The results seen with respect to dynamic range compression, edge enhancement,
and local gain adaptation seem to simultaneously address the issues of coding, enhancement and feature extraction in image processing and warrant continued investigation of
nonlinear models of lateral inhibition.

Figure 4.11 The mean luminance of the original image scaled to the range [0 to 255] gray.

Figure 4.12 Log-scale of the original image scaled to the range 0 to 255 gray.

The models have been formulated in a discrete signal processing framework and their behavior as digital image processing systems has been evaluated. The simplified model demonstrates the major characteristics of interest for this class of model and allows for efficient exact and experimentation of these characteristics as the model parameters change. The results even with respect to dynamic range compression, edge enhancement, and local gain adaptation seem to simultaneously address the needs of cueing, enhancement and feature extraction in image processing and represent combined realization of nonlinear models of neural behavior.

5

Modeling the Processing and Perception of Visual Motion

Abdesselam Bouzerdoum and Robert B. Pinter

5.1 INTRODUCTION

The processing of motion information is a fundamental and elementary function in biological visual systems. Perception of depth, segregation of objects, discrimination of figure from the ground, and detection of moving objects are a few tasks among many others that rely on visual motion perception; see (Nakayama 1985) for a review. Sophisticated mechanisms for extracting and utilizing motion information exist even in simple animals. For example, the ordinary housefly can separate a moving object from its surround on the basis of motion information alone (Egelhaaf 1985; Reichardt 1986; Reichardt and Poggio 1979). The frog does not recognize a dead insect as food; however, the frog has efficient "bug-detection" mechanisms that respond selectively to small, dark objects moving across its visual field (Lettvin et al. 1959).

However, the idea that motion is a basic sensory dimension and not just a perception compounded from sensations of time and space did not arise until 1875 when Exner discovered that motion perception is more precise than the perception of time-order. He presented two electric sparks, separated in both space and time, to an observer who had to judge the order in which the sparks were flashed. Exner found that when the time interval separating the two flashes was less than 45 ms, the observer could not correctly identify their order. But when the two sparks were brought close to each other, the subject reported seeing the first spark move from its location to that of the second spark even when the time intervals between flashes were as short as 14 ms, that is, shorter than the perceptible time interval of the first experiment. Exner concluded that motion must therefore be a sensation in its own right.

Exner's discovery was confirmed 37 years later by Wertheimer (1912). Since then motion has been the subject of intensive studies comprising psychophysical, neurophysiological, behavioral, and computational. The Gestalt psychologists labeled this phenomenon of two brief flashes of light creating an illusion of motion under the appropriate spatial and temporal separations, *apparent motion*. The Gestalt psychologists may be credited with popularizing the study of apparent motion; however, their attempts to explain the brain mechanisms underlying motion perception were premature. Significant progress in this regard came with the discovery of neurons in the frog retina that respond selectively to small dark moving objects (Lettvin et al. 1959). Since then motion sensitive neurons were identified in several other species: pigeons (Maturana and Frenk 1963); rabbits (Barlow et al. 1964); ground squirrels (Michael 1968); flies (Bishop et al. 1968); cats (Hubel and Wiesel 1959); monkeys (Schiller et al. 1976); and others.

A large number of motion theories have been proposed which lead to various classes of algorithms for encoding direction and velocity. Surprisingly, in spite of fundamental differences many theories share some overall organizational aspects of the motion system. Evidence from recent psychophysical experiments suggests that in the human visual system there are at least two distinct mechanisms for analyzing motion perception (Anstis 1980; Braddick 1980, 1974). Braddick (1974) termed them *short-range* and *long-range processes*. He further reported that the short-range process operates on continuous motion or discrete motion but over rather short spatial separations (at most 10–15' of visual arc) and short time intervals (up to 60–100 ms). The long-range process, however, operates in classical apparent motion situations with larger spatial displacements (up to several degrees of visual arc) and longer time intervals (up to 500 ms). Furthermore, the short-range is a peripheral process that involves low-level kinds of visual information, and is likely mediated by directionally selective neural units. The long-range process, on the other hand, appears to be more central with perhaps a cognitive component, thus it requires somewhat higher level forms of visual information.

Computationally, there exist mainly two sets of approaches to motion analysis, namely the *feature-matching schemes* and the *intensity-based schemes*. The former schemes may play a central role in the processing of *tasks of integration*, which require a combination of information over time and cannot be solved using solely instantaneous measurements. Ullman (1979a, 1979b) has shown that the recovery of three-dimensional structure is a task of integration. The latter schemes, on the other hand, are useful for processing *tasks of separation*, like discrimination of moving objects from one another and from the background. These tasks can in principle be solved by using instantaneous measurements, such as position and its time derivative in the image. In the human visual system, it seems that the short-range process is an intensity-based scheme; whereas the long-range process, which has been suggested to be crucial to the recovery of structure from motion, (Petersik 1980), is a feature-matching scheme.

These approaches to motion detection and measurement give rise to different computational problems in biological and machine vision systems. To illustrate different solutions to these problems, some models from feature-matching schemes and from intensity-based schemes are briefly introduced in the succeeding two sections. Several of the models were developed in artificial intelligence and machine vision; hence, they may be unfit for modeling motion perception in biological systems. *Shunting inhibition-based* models of direction selectivity are presented in Section 5.4. In particular, a shunting inhibitory motion detector that belongs to the family of intensity-based schemes is developed and some of its response characteristics are discussed.

5.2 FEATURE-MATCHING SCHEMES

Motion can be described by establishing an explicit correspondence over time, between elements representing the same physical feature in the same image. In feature-matching schemes, also known as *token-matching schemes*, motion measurement requires locating identifiable features such as edges, corners, blobs, or regions in the changing image and matching them over time. The visual input may consist of stimuli presented as sequences of discrete frames (e.g., cine film), in which case the rate of translation can be inferred by computing the ratio $(\Delta S/\Delta t)$ of ΔS, the distance traveled, to Δt, the time between successive frames.

The problem of matching a particular feature in an image frame with the right feature in subsequent frames, is known as the *correspondence problem*. Such a problem can be tackled not only at the level of simple features such as points, blobs, or small image segments, but also at the level of complex features like surfaces, structured forms, or even whole objects. Solving the correspondence problem—in the sense of finding the best match—usually requires some optimality criteria. Several techniques that differ in optimality criteria and in the level at which correspondence is established (i.e., the degree of preprocessing and the complexity of participating features), have been suggested for motion detection and measurement.

A wide variety of criterion functions, ranging from simple cross-correlation (Gennery 1979; Yakimovsky and Cunningham 1978) to complex and sophisticated graph-matching procedures (Jacobus et al. 1980), has been utilized in the framework of computer vision research. In the same vein matching techniques were also diversified. For example, Yachida et al. (1981), and Martin and Aggarwal (1979) relied on image segments, while Williams (1980) used full object surfaces. Alternatively, Potter (1975, 1977), and Barnard and Thompson (1980) used derived feature points and adaptive template matching to solve the correspondence problem. The algorithm presented by Barnard and Thompson breaks the matching task into two steps. First, feature points are identified by searching each frame for areas of high variability in several directions. Then by successive approximation, match probabilities are adjusted for pairs of match points. If the algorithm converges, correct and incorrect matches will have high and low probabilities respectively. The problem is that the algorithm may not converge, and there is no simple way to check whether or not it has indeed converged. Lawton (1983) also developed a procedure that consists basically of two steps: feature extraction and search. First features consisting of small image areas, which may correspond to parts of environmental objects, are extracted. The search process then finds the direction of translational motion. The right direction of translation determines the corresponding image displacement paths (curve along which an image feature point is moving) for which a measure of feature mismatch is minimized.

In dynamic scenes, approaches to correspondence can be partitioned into two classes based on the rigidity of objects and continuity of motion, which are known to play important roles in the perception of apparent motion (Ramachandran and Anstis 1986). Several approaches emphasizing rigidity of objects were developed lately, one of them is the relaxation algorithm of Barnard and Thompson (1980) discussed previously. These algorithms translate the rigidity property into the spatial smoothness of the disparity vector field which maps the feature points of one frame onto the feature points of the next frame. Although the spatial smoothness assumption may result in poor performance when dynamic scenes contain several moving objects, the rigidity-based approaches offer the advantage of equally applying the matching procedure to many other situations such as in stereo matching. The second class of algorithms adopts the motion continuity property by assuming temporal smoothness of disparity vectors and allows the rigidity assumption to be relaxed. This is very useful in dynamic scenes where more than two frames are to be analyzed because simultaneous matching over all frames is possible. Approaches based on continuity of motion are expected to perform much more effectively with multiple moving object scenes owing to the fact that the rigidity assumption can be relaxed. The relaxation algorithm proposed by Sethi et al. (1988) is a good example of approaches exploiting motion continuity to establish correspondence. Using the hypothesis of smoothness of motion in time, the algorithm updates iteratively the initial match values in order to obtain smooth trajectories over space and time indicating the feature point matches. This algorithm was developed to deal with the problem of inconsistency of feature points, where their number changes drastically from frame to frame for one reason or another.

The apparent motion phenomenon indicates the ability of the human visual system to establish motion by solving the correspondence problem (Braddick 1974; Julesz 1971; Kolers 1972; Wertheimer 1912). However, the correspondence process is far from being completely understood in human vision despite the fact that it has been extensively investigated (Anstis 1980; Frisby 1972; Ramachandran and Anstis 1986; Ullman 1980, 1978). Psychophysical evidence indicates that many factors serve as cues for detecting correspondence. These factors include rigidity of objects, continuity of motion, proximity and similarity effects, visual texture, and stimulus orientation (see Ramachandran and Anstis 1986). Ullman (1980) suggested that interactions between the short-range and long-range processes are likely to play a part in determining perceived motion, thus an integration of intensity-based and feature-matching schemes may be necessary for coding visual motion.

Ambiguity in matching.   Feature-based models have difficulty making predictions because of the familiar problem: what is to be considered a feature, and how such a feature may be optimally detected. In addition, ambiguity in matching (i.e., inability to identify a feature's correct match among other possible matches), constitutes a major hurdle in solving the correspondence problem for both physical and biological systems. The task of developing the best combination of ambiguity avoidance and resolution has recently become a fundamental research issue. When the human visual system is faced with ambiguity, it determines correspondence by extracting salient features from images and also limiting motions to those consistent with palpable universal laws of matter and motion.

5.3 INTENSITY-BASED SCHEMES

The coding of motion, which may take place at different stages in the processing of an image, could be determined at the lowest stage by low-level properties of stimuli such as local changes in light intensity. Motion detection models that are based upon raw intensity values can be classified as *global* or as *local* depending on how the model analyzes an image frame. Global models analyze the entire frame or a significant portion of a frame. On the other hand, local models infer motion of larger areas by combining the outputs from a large number of units (i.e., movement detectors); each unit computes the direction of motion within a small area.

5.3.1 Global Models

Two distinct methods have mainly been advanced for analyzing motion on a point-by-point basis from an entire frame: *Cross-correlation* and *differencing methods*. These two methods have been developed so as to describe and compute similarities and differences respectively between two frames. Other techniques, such as *spatial phase analysis* or *transform methods* e.g., Fourier transform (Huang 1981) and Hadamard transform (Fu and Chang 1989; Lai and Chang 1988), have also been used but not so extensively as the aforementioned methods.

Cross-Correlation Methods.   Motion can be extracted by means of similarity measures between successive frames. One of the simplest ways to achieve this is to find the amount of spatial shift, ΔS, that maximizes the gray-level cross-correlation between the two frames. Cross-correlation techniques are widely used in different applications. For example, they have been used for cloud tracking and cloud motion estimation from satellite photographs (Leese et al. 1970; Smith and Phillips 1972), as well as for traffic control (Wolferts 1974). Furthermore, cross-correlation has been suggested as a model for interpreting human visual motion perception (Bell and Lappin 1973; Petersik et al. 1978).

Optical Correlation.   Early optical correlation-based models for motion processing have originated from the practical need to measure the optical speed of moving objects for specialized applications (e.g., speed of the ground from an aircraft taking aerial photographs) (Agar and Blythe 1968; Ator 1966, 1963). These models, however, do not measure the direction of motion. Optical correlation methods include autocorrelation and spatial frequency filtering techniques.

1. **Autocorrelation** The autocorrelation process is carried through continuous computation of the correlation existing between a signal received in real time and a signal previously obtained from the same source. The general implementation method for the autocorrelation approach is known as the *two-slit correlator* (Ator

1966). This device consists of two photocells viewing the same object at different times. The signal from one photocell is continuously recorded on a magnetic tape. The recorded and real signals are then correlated to determine the time delay for which the two signals exhibit maximum similarity.

2. **Spatial Frequency Filter** The principal component of a spatial frequency filter is a parallel-slit reticle with regular periodic arrangement placed over a photosensor. The filter is tuned to a single spatial frequency which is the frequency of the parallel-slit pattern. The light energy transmitted by the filter is modulated in time according to the spatial frequency of the reticle and the velocity of the image. Thus, the signal produced from sensing this modulation is of sinusoidal type with a temporal frequency directly proportional to the transverse component of the image velocity (Agar and Blythe 1968; van Doorn and Koenderink 1983; Ator 1966). The transverse velocity, V_T, can be calculated by dividing the temporal frequency, f_t, of the photosensor output by the spatial frequency, f_s, of the reticle,

$$V_T = \frac{f_t}{f_s}. \tag{5.1}$$

However, the longitudinal component and the direction of the velocity vector remain undetermined. The spatial frequency filtering method is referred to as *optical correlation processing* in the sense that the filter's transmittance function is cross-correlated with the image's light flux distribution function.

Differencing Methods.   Related to cross-correlation techniques are differencing techniques, where significant changes in image intensity can be efficiently detected by subtracting one image from the other, and thresholding the result so that differences caused by noise are discarded. When an image frame is subtracted from its predecessor, moving image components show double-images in the difference picture, while stationary components cancel out and disappear. Unfortunately, due to the thresholding operation slow moving components are eliminated as well. Several techniques have been developed to generate the difference picture. They range from simple subtraction to a likelihood ratio based on the second-order statistics of small regions. In computer vision differencing techniques have been utilized to extract moving object images (Jain and Nagel 1979) and to segment dynamic scenes (Jain et al. 1979); together with cross-correlation they were used for the measurement of motion (Jain et al. 1977; Nagel 1978).

Since the subtractive process detects only the presence of a change, discrimination of direction and velocity of motion requires an appropriate spatial shift of one frame over the other prior to subtraction. If the image contains components with different motion characteristics (e.g., different velocities), the number of shift operations required increases considerably and augments with it the computational load. There are many deficiencies of global models in general. They are designed to make predictions about stimuli presented as sequences of discrete frames; however, not all stimuli fit such description. But the major problem is that global models assume the entire image (or a significant fraction of it) moves as a whole between two successive frames. As a result, global models can usually make predictions about simple stimuli such as a moving bar, but they may ambiguously sense motion when presented with an image of independently moving objects.

5.3.2 Local Models

A large number of motion-detecting models that explicitly use local changes of intensity over space and time to infer motion have been suggested. These models of local motion computation can be categorized into: *spatio-temporal gradient schemes*, *spatio-temporal energy models* and *delay-and-compare schemes*.

Spatio-Temporal Gradient Schemes.   The optical flow field, or the velocity field that represents the motion of objects across an image, can be computed from local temporal and spatial variations of image intensity. The gradient techniques are one way of calculating the optic flow from local information (Fennema and Thompson 1979; Horn and Schunck 1981; Marr and Ullman 1981). They are based upon the common assumption of constant brightness, that is, the observed intensity on the image plane is constant over time. Therefore, any changes in intensity must be induced by motion. This assumption (to first order) implies the gradient constraint formally stated in Eq. (5.2), whose detailed derivation was given by Horn and Schunck (1981), see also (Schunck 1985, 1986).

$$\nabla \vec{I} \cdot \vec{V} = -\frac{\partial I}{\partial t} \qquad (5.2)$$

This equation, known as the brightness constancy equation or the image flow constraint equation, relates the temporal derivative of the intensity function ($\partial I/\partial t$) at a point and the spatial gradient $\nabla \vec{I}$ in the neighborhood of the point to the instantaneous velocity $\vec{V}$ at that point in the image. This equation in general is not valid because it applies only when the change in image intensity at each point in the image plane is entirely due to translational motion and the image is smooth except at a finite number of discontinuities, (for a more detailed discussion see, e.g., Schunck, 1985, 1986). However, it has been shown that for sufficiently textured objects, the brightness constancy equation is approximately true for a large number of reflectance functions (Verri and Poggio 1987).

Equation (5.2) constrains the velocity vector to lie on a line in velocity space perpendicular to the intensity gradient vector, but does not determine it uniquely. Thus, the gradient constraint merely yields the component of the velocity vector in the gradient direction. So, using this equation as the only constraint to solve for the optical flow results in an underconstrained problem and hence ambiguity in motion. This ambiguity is commonly referred to as the aperture problem, which arises when a long straight moving edge is observed through a narrow aperture (Ullman 1979a; Marr and Ullman 1981; Hildreth 1984). The ambiguity, however, can be resolved by taking the spatial derivatives of the gradient constraint equation except in the case of special surfaces that possess a singular Hessian matrix, like hyperboloids (Uras et al. 1988; Reichardt et al. 1988; Horn 1987).

Other investigators have also used second order spatial and spatio-temporal derivatives to solve for the optical flow (Nagel 1983, 1986, 1987; Tretiak and Pastor 1984).

But in most algorithms proposed for recovering the missing optic flow component, a second constraint was introduced. Three ways to express the additional constraint have been suggested: local optimization, global optimization, and clustering. Both local optimization and clustering methods make use of the additional constraint of constant flow velocity within a small region of the image to arrive at a unique solution. In local optimization (Kearney et al. 1987; Thompson and Barnard 1981), a set of gradient constraint lines from a small neighborhood is solved as a system of linear equations (two constraint lines are sufficient to determine a unique velocity vector). Clustering methods on the other hand operate globally, seeking groups of points having the same velocity vector (Fennema and Thompson 1979). Because the constant flow velocity constraint stems from the assumption of uniform rigid body translation, local optimization and clustering techniques cannot deal with objects involving rotation. In contrast, global optimization techniques deal more efficiently with rotating objects by minimizing an error function based on the gradient constraint and an additional assumption of local smoothness of the optical flow field.

Horn and Schunck (1981) have formulated the constraint of smoothness as the gradient magnitude square of the intensity function. Other measures of smoothness include the sum of the laplacian of the flow components (Horn and Schunck 1981; Nagel 1983, 1987) and the variations of the velocity vector (Hildreth 1984). The general smoothness requirement of Horn and Schunck, however, creates difficulties since it forces the optical flow to vary smoothly even across the image of occluding contours. To cope with this difficulty, Nagel (1983, 1986, 1987) suggested the "oriented smoothness" constraint which requires smooth variations of the optical flow only in directions with small or no variation of image intensity, such as the direction perpendicular to the gradient.

Zero-Crossing Model.   Marr and Ullman have used spatial filtering to provide a more plausible realization of Eq. (5.2). They have suggested that initial motion measurements in the human visual system take place at the locations of sharp intensity changes. To detect these intensity changes, Marr and Hildreth (1980) proposed that an optimal operator for the initial filtering of the image is the *laplacian of gaussian* (LOG), whose shape may be approximated by the difference of two gaussian distributions (DOG). The zero-crossings in an image filtered through the LOG operator correspond to sharp intensity changes in the image. Accordingly, Marr and Ullman (1981) proposed an algorithm where initial motion measurements are made only at the locations of zero-crossings, using a mechanism that combines the temporal derivative and the spatial gradient of the filtered image. The proposed zero-crossing model computes very economically the direction of motion by considering only the signs of the spatial and temporal derivatives of the filtered image. The magnitude of the spatial and temporal gradients can be further used to encode the speed of movement (Harris 1986). A possible neural model that computes the spatial and temporal derivatives was suggested by Richter and Ullman (1982).

The theory of Marr and his collaborators suggests that the retinal ganglion X-cells perform the spatial filtering operation while a class of simple cortical cells may be involved with the zero-crossing detection. In addition, the Y-cells are concerned with the temporal derivative of the filtered image. The hypothesized role of filtering for X-cells is in general agreement with physiological evidence (the receptive field profile of these cells can be approximated by a DOG); however, it is still not clear whether the cortical cells are indeed concerned with the detection and representation of zero-crossings (Richter and Ullman 1986). Y-cells are unlikely to transmit the sign of the temporal derivative of the filtered image because they are fed by rectifying subunits. Although this suggests that Marr and Ullman's model of motion detection does not agree with electrophysiological data on motion detection in biological systems, its computational aspects still stand as a new theoretical contribution.

Spatio-Temporal Energy Models.   The basic principle on which energy models operate to extract motion information is that translation at constant velocity shears the spectrum of a stationary image into a plane in the spatiotemporal frequency domain. To see this, consider a two-dimensional (2-D) intensity pattern $I_0(x, y)$ translating at constant velocity $\vec{V} = (v_x, v_y)$, that is, the spatiotemporal intensity function is given by

$$I(x, y, t) = I_0(x - v_x t, y - v_y t).$$

The three-dimensional Fourier transform of this function is related to that of $I_0(x, y)$ by the following equation:

$$\mathcal{I}(\omega_x, \omega_y, \omega_t) = \mathcal{I}_0(\omega_x, \omega_y)\delta_p(\omega_t + v_x\omega_x + v_y\omega_y),$$

where δ_p represents the plane impulse function; it is zero everywhere except on the plane:

$$\omega_t + v_x\omega_x + v_y\omega_y = 0. \tag{5.3}$$

Therefore, the spectrum of the moving 2-D pattern lies in an oblique plane through the origin whose angle with the plane $\omega_t = 0$ is determined by the speed $|\vec{V}|$, and whose orientation about the ω_t-axis is determined by the direction of the velocity vector $\vec{V}$. Analogously, the spectrum of a 1-D pattern translating in the image plane with constant velocity lies along a line in the frequency space.

Motion in general cannot be readily represented by a plane in the frequency domain; but if only small spatiotemporal apertures are considered, motion can be approximated and the velocity may be computed by finding the plane in which the power resides. Energy models extract motion information by detecting the oriented spatiotemporal energy. This can be accomplished using a quadrature pair of spatiotemporally oriented filters, such as the Gabor energy filters (Adelson and Bergen 1985), or Hilbert filters (Watson and Ahumada 1985). Spatiotemporally oriented filters are tuned to particular spatiotemporal frequencies, but are not velocity-selective mechanisms. A single such mechanism is unable to distinguish between variations in the stimulus spatiotemporal frequency contents or variations in its contrast. However, an unambiguous velocity estimate may be obtained by combining the responses of a collection of spatiotemporally oriented filters (Heeger 1988; Watson and Ahumada 1985). This helps resolve the ambiguity resulting from the aperture problem, which occurs when all the spatial frequencies of a moving image have the same orientation. The spectrum of such an image lies along a straight line through the origin. Since the line is contained within an infinite number of planes, it is consistent with an infinite number of velocities.

Delay-and-Compare Schemes.   It is widely known that in order to recover a three-dimensional object structure from motion, the velocity must be measured very accurately; whereas to respond quickly to a moving object, motion must be detected but not necessarily measured (Clocksin 1980; Longuet-Higgins and Prazdny 1980; Prazdny 1980; Ullman 1981, 1979a, 1979b). Therefore, it may be sufficient for some visual tasks to compute only certain properties of the velocity field. In fact, many biological visual systems, (for example, insects), have not evolved much beyond the ability to separate the direction of discontinuity in the optical flow field. Figure 5.1a depicts the simplest system that responds to motion, namely the *elementary movement detector (EMD)*. It consists of two receptor regions, R_1 and R_2, separated by a distance ΔS. A moving stimulus is sampled by the receptors R_1 and R_2, and the signal from R_1, after being delayed, is compared with that from R_2. If an object is moving from left to right at constant speed V, then with the "appropriate" time delay ($\tau = \Delta t = \Delta S/V$) the signals to be compared are the same which indicate rightward motion. However, if the object is moving in the opposite direction, the signal from R_2 arrives to the comparator before the signal from R_1 does. This means the signals to be compared are different and so no motion will be signaled. The EMD is, thus, directionally selective: it has a *preferred* direction (left to right), and a *null* direction (right to left).

A movement detector which responds to either motion (leftward or rightward), comprises two mirror-symmetrical subunits (EMDs) tuned to motion in opposite directions (Fig. 5.1b). The detector output is formed by subtracting the response of the left subunit from that of the right subunit. Consequently, the detector response will be positive, implying rightward motion, when the output of the right subunit exceeds the output of the left subunit, and vice versa. While the most prominent feature of this movement detector is its simplicity, its major drawback is achieving an "appropriate" time delay that matches every velocity of interest. The "appropriate" delay is a crucial part of the detector because even equipped with the right method of comparison, the EMD does not respond unless the signals from the two sampling channels arrive simultaneously at the comparator.

The movement detector just described is the prototype of a large class of motion computation models known as the *delay-and-compare schemes*, most of which were proposed to explain data from psychophysical (Foster 1971; Wilson 1985; van Santen and Sperling 1985), neurophysiological (Barlow and Levick 1965; van Doorn and Koenderink 1976; Ruff et al. 1987), and behavioral experiments (Hassenstein and Reichardt 1956; Reichardt 1961; Reichardt 1986; Reichardt and Egelhaaf 1988). Different models differ in the way the delay-and-compare operation is implemented. Some models use simple delay filters (pure time delays or low pass filters) followed by simple logical or correlation operations (Barlow-Levick and Reichardt correlation models), whereas others utilize sophisticated spatiotemporal filtering operations to implement the "appropriate" delay and method of comparison (Wilson 1985). Here, only Barlow-Levick and Hassenstein-Reichardt early models are discussed.

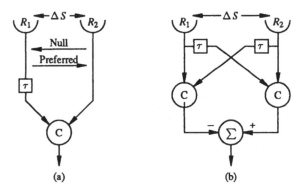

Figure 5.1   (a) Elementary Movement Detector (EMD). The outputs from two adjacent detectors R_1 and R_2, separated by a distance ΔS, are connected to a comparator ⓒ, one via the appropriate time delay (box τ). (b) Two mirror-symmetrical EMDs sharing the neighboring input channels which sample the visual field, form a motion detector whose final output is given by the difference between the EMDs outputs.

Barlow-Levick Model.   There is no doubt that biological visual systems possess neurons that respond selectively to motion in a particular direction termed the *preferred* direction. Movement in this direction evokes a vigorous response, whereas movement in the opposite, or *null,* direction yields no significant response. Directionally selective cells can be found in both the retina and higher level units (Barlow et al. 1964; Hausen 1981; Hubel and Wiesel 1959; Schiller et al. 1976). To explain direction selectivity of ganglion cells in the rabbit retina, Barlow and Levick (1965) proposed two conceptually different gating schemes. One selects the preferred stimulus by detecting a specific conjunction of excitation using a logical AND gate as a comparator. The other works by rejecting the null stimulus by a veto operation using an AND-NOT gate instead of the comparator. These models, however, still need the "appropriate" time delay for optimum direction discrimination to happen.   Both rapid and slow motions in either direction trigger a response in the inhibitory (AND-NOT) model and give no response in the excitatory one (AND-gate model).

Experimental evidence in many cases favors the *veto* scheme, for it was demonstrated that the mechanism mediating direction selectivity is of the inhibitory (or veto) type. In their study, Barlow and Levick (1965) proposed that inhibition is triggered selectively in a way that at each subunit a delayed inhibitory mechanism vetoes the excitatory response in the null direction, but appears too late to cancel the response in the preferred direction. Their ideas have since received further extensive experimental support. For example, Michael (1968) found that when simultaneously moving two dots in opposite directions, the spot moving in the null direction inhibited the response to the spot moving in the preferred direction, but there was no interaction when the distance separating the two dots was appreciable. Other investigators have reported that the inhibitory neurotransmitter GABA (*Gamma-Amino-Butyric Acid*) is responsible for the veto mechanism in directionally sensitive cells. By blocking the GABA action with drugs, the cells yielded equal responses to motions in both the preferred and null directions (Schmid and Bülthoff 1988; Ariel and Adolph 1985; Ariel and Daw 1982; Caldwell et al. 1978; Sillito 1975).

Correlation Model.   Barlow and Levick's model is the first directionally selective neurophysiological model, but an earlier one was a mathematical model formulated by Hassenstein and Reichardt in 1956 to account for the characteristics of insect optomotor response. In this model the delay operation was implemented by introducing into each leg of the EMD a linear time-invariant low-pass filter—the delay filter. Knowing that a linear temporal filter delays different modulation functions by different amounts, Hassenstein and Reichardt used two filters with short and long time constants, yielding a near constant phase asymmetry between the EMD legs over a reasonable range of temporal frequencies. Others have substituted a high-pass filter for one of the low-pass ones in order to generate the same phase lag between the EMD legs for an extended range of modulation rates (Buchner 1976; Zaagman et al. 1978). Moreover, the comparison operation was implemented by multiplication followed by infinite time averaging, that is, first-order correlation. Multiplication has been shown to be indeed the mechanism mediating movement detection in the visual system of insects (Hassenstein and Reichardt 1956; Reichardt 1961; Buchner 1976; Zaagman et al. 1978). Different variants of the correlation model were also proposed in various contexts by Thorson (1966), Foster (1971), and van Santen and Sperling (1984). Movement detectors of the correlation type received further support when Marmarelis and McCann (1973), studying movement detection in class II neurons of the fly's visual ganglia, found that nonlinearities higher than second order do not contribute significantly to the response. Furthermore, Poggio and Reichardt (1973) demonstrated that if nonlinearities higher than second order are negligible, then any movement-detecting system is equivalent to the correlation model provided that the infinite time average of the output is taken.

Spatial and Temporal Aliasing.   The original Hassenstein and Reichardt correlation model is susceptible to some form of aliasing. Since the model employs spatial sampling, it is necessarily subject to the limitations on spatial resolution set by the Shannon sampling theorem. This can be clearly seen if one considers the response of the system to moving sine waves. Let a drifting sine-wave grating be defined as

$$I(s,t) = I_0 + \Delta I \cos(2\pi\mu f_s s - 2\pi f_t t + \varphi), \qquad (5.4)$$

where I_0 is the mean intensity, ΔI is the modulation amplitude, μ is the direction of movement (-1 for leftward motion, $+1$ for right ward motion), f_s and f_t are the grating spatial and temporal frequencies respectively, and φ is the intial phase. If $\Delta\varphi$ designates the difference in phase between the EMD legs, then it can be easily shown that the mean response of a correlation-type EMD, $\overline{r(t)}$,[1] is such that

$$\overline{r(t)} \propto \cos(\Delta\varphi - 2\pi\mu f_s \Delta S). \qquad (5.5)$$

---

[1] The bar indicates that an infinite time-averaging operation has been taken.

If $\Delta\varphi = \pi/2$ for all f_t, then $\overline{r(t)}$ is proportional to $\sin(2\pi\mu f_s \Delta S)$, or

$$\overline{r(t)} \propto \mu \sin(2\pi f_s \Delta S). \qquad (5.6)$$

The sign of the mean response depends on μ, hence the response is fully directional. Now let $\overline{d(t)}$ be the response of the full correlation motion detector, that is, the detector in Fig. 5.1b with correlation-type EMDs. It follows from Eq. (5.5) that

$$\overline{d(t)} \propto \sin(\Delta\varphi)\sin(2\pi\mu f_s \Delta S). \qquad (5.7)$$

The sign of the mean response again depends on μ. In accordance with the sampling theorem, if the distance between the input channels ΔS is less than half the grating wavelength λ (i.e., $\Delta S < \lambda/2$), $\sin(2\pi\mu f_s \Delta S)$ then has the same sign as μ. However, as the spatial frequency increases λ decreases; when $(\lambda/2 < \Delta S < \lambda)$, the sign of $\sin(2\pi\mu f_s \Delta S)$ becomes opposite that of μ. Direction selectivity then reverses sign and the detector signals the wrong direction. This problem is known as *spatial aliasing*. Temporal aliasing, on the other hand, arises when $\sin(\Delta\varphi)$ changes, leading to incorrect direction of motion. The aliasing problem may, however, be eliminated by including spatial-frequency-tuned receptive fields in the input channels. For more on what conditions spatial and temporal filters must have to prevent aliasing (see e.g., van Santen and Sperling 1985, 1984).

5.4 SHUNTING LATERAL INHIBITORY MODELS

Barlow-Levick and Hassenstein-Reichardt directionally selective models differ in their approach to solving the motion detection problem. The former, based on physiological study, attempts to simulate the neural circuitry underlying direction selectivity in the ganglion cells; whereas the latter, based on behavioral analysis, seeks to establish the functional principles of motion detection as inferred from the average optomotor response of insects. Although these two early models differ in some respects, they are similar in many others. Both models operate on the same principle of a nonlinear asymmetric interaction between channels from two adjacent receptor regions (see Fig. 5.1). In fact, any system whose average response is directional must consist of at least two nonlinearly-interacting asymmetric channels (Buchner 1976; Poggio and Reichardt 1973). Hassenstein and Reichardt (1956) utilized multiplication as the nonlinear interaction between the asymmetric channels. Since then overwhelming evidence has been accumulated that the nonlinear interaction mediating motion detection is of the multiplicative type (Buchner 1976; Egelhaaf and Reichardt 1987; Franceschini et al. 1989; Reichardt 1986, 1961).

At the single cell level, the nonlinear interaction often turns out to be inhibitory (Barlow and Levick 1965; Michael 1968; Schmid and Bülthoff 1988; Ariel and Daw 1982; Sillito 1975). Barlow and Levick did not explicitly specify the nature of the inhibitory interaction; however, Torre and Poggio (1978) suggested *feed-forward shunting inhibition* (Furman and Frischkopf 1964) as a plausible biophysical mechanism for the actual neural implementation of the multiplication-like interaction.[2] They postulated

---

[2]Shunting inhibition has been described as a multiplicative operation for small enough inputs (Thorson, 1966; Torre and Poggio, 1978).

that the two asymmetric input channels serve as two closely adjacent excitatory and inhibitory synaptic inputs to a patch of passive membrane. Then, they showed that shunting inhibition, or inhibitory synapse with a reversal potential close to the resting potential of the cell, can subserve directional selectivity as described by Barlow and Levick (1965).

In the nervous system it is likely that *recurrent shunting inhibition* is used, where outputs *feed-back* to input nodes. The feed-back is via the biophysical mechanism underlying shunting inhibition: modulation of the shunting conductance by voltages of neighboring cells or cell subunits. Thus, shunting inhibition translates into multiplicative interaction between voltage functions of neighboring neurons (Bouzerdoum and Pinter 1989b; Pinter 1983). This sort of shunting inhibition is referred to as *multiplicative lateral inhibition (MLI)*. Multiplicative lateral inhibitory neural networks (MLINNs) are described by a system of coupled, nonlinear differential equations of the form:

$$\frac{de_i}{dt} = L_i(t) - k_i e_i \left[1 + \sum_{j \neq i} k_{ij} f_{ij}(e_j) \right], \ i = 1, 2, \ldots, \quad (5.8)$$

where e_i describes the deviation of the membrane voltage from the resting potential; $L_i(t) > 0$, is the external input to the ith node or neuron; $k_i > 0$, is the rate constant for the ith neuron; $k_{ij} > 0$, represents the synaptic efficacy (or coupling strength) of the jth neuron potential transmitted to the ith neuron; and f_{ij} is the lateral inhibition function: it is positive for positive arguments and represents the dependence of the shunting conductance in the ith neuron on the voltage of the jth neuron.

5.4.1 Directional Selectivity with MLINNs

The property of directional selectivity can be accounted for by introducing asymmetrical coupling among elements of MLINNs. It has been shown that unidirectional MLINNs, or networks with one-sided coupling such as those described by Eq. (5.9), are stable (Bouzerdoum and Pinter 1989b) and possess a directional response to motion in opposite directions, that is, the response depends markedly on the direction of motion (Bouzerdoum and Pinter 1989b, 1990b). This ability of unidirectional MLINNs to discriminate between motions in reverse directions is demonstrated in Fig. 5.2, which depicts the response of the second neuron to a bar moving in opposite directions.

$$\frac{de_i}{dt} = L_i(t) - k_i e_i \left[1 + \sum_{j < i} k_{ij} f_{ij}(e_j) \right], \ i = 1, 2, \ldots \quad (5.9)$$

In attempt to quantify the directional selectivity of unidirectional MLINNs, define the percent gain, $G(\%)$, as the ratio of the difference to the sum of incremental sensitivities of responses to motion in opposite directions, that is

$$G(\%) = 100 \left(\frac{Gr - Gl}{Gr + Gl} \right),$$

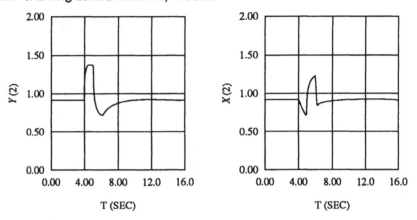

Figure 5.2 Response of the second neuron of a unidirectional MLINN to a bar moving in opposite directions: The response on the left is for leftward motion and the one on the right is for rightward motion. The network is described by Eq. (5.9) with lateral inhibition function $f(e_j) = e_j$, $L_2(t) = L_0 + \frac{1}{2}L_0[u(t-5) - u(t-6)]$, and mean luminance $L_0 = 10$ (arbitrary units). Most units of this array possess asymmetrical responses to motion in two opposite directions.

where Gr is the incremental sensitivity (gain) in response to rightward motion (i.e., motion in the coupling direction for networks of Eq. (5.9)) and Gl is the incremental sensitivity for motion in the opposite direction. These percent gains vary with mean luminance, speed and size of moving objects, and the order of coupling in the network (i.e., the number of units with which a single neuron is coupled) (Bouzerdoum and Pinter 1989b, 1990b). The percent gains for a unidirectional MLINN with nearest neighbor coupling only, are presented in Fig. 5.3 as functions of node number in the array. Note that the percent gains in this figure increase with mean luminance, L_0, and for each mean luminance level, they approach a constant value towards the end of the array.

For infinitely long arrays, the percent gains have a limit, which depends on L_0, as the node number approaches infinity (proof is in Bouzerdoum's personal notes). Let this limit be the percent directionality index of the network, or

$$Directionality\ Index(\%) = \lim_{i \to \infty} G_i(\%)$$

where $G_i(\%)$ is the percent gain for the ith node. Figure 5.4 shows the directionality indices for 3 unidirectional MLINNs having different coupling orders. The directionality index of each of these networks increases as L_0 increases but saturates at high values of L_0. Figure 5.4 also shows that the directionality index decreases as the order of coupling increases. But this is not true in general, because the size of the moving object also plays a role in determining the directionality of the network. For example, in Fig. 5.5 (on page 64) the directionality index increases with the coupling order for relatively large bars. Another factor which affects the directional selectivity of a network is the speed of the

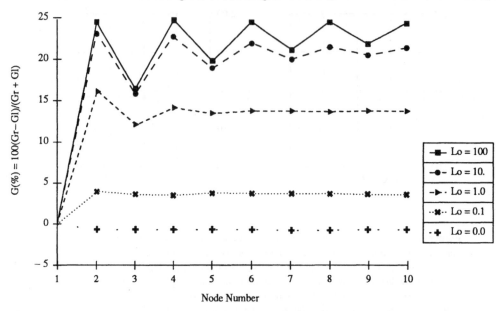

Figure 5.3 Percent gains for a unidirectional MLINN in response to moving bright bar. $\dot{e}_i = L_i(t) - e_i(1 + e_{i-1})$, speed = 0.2 comp/sec, bar size = 1 comp. Here, 1 compartment (comp) = spatial (or angular) separation between neighboring nodes.

moving object; the directional selectivity usually degrades as the speed of the moving object increases (see Fig. 5.6 on page 65).

5.4.2 Multiplicative Inhibitory Motion Detectors

Since unidirectional MLINNs respond differentially to motion in opposite directions, they may be used to form motion detectors of the delay-and-compare scheme-type. These motion detectors were named the *multiplicative inhibitory motion detectors (MIMDs)* (Bouzerdoum and Pinter 1989a). The remainder of this section is devoted to the properties of such motion detectors.

Figure 5.7 on page 66 illustrates a MIMD composed of two mirror-symmetrical unidirectional MLINNs each comprising two nodes. The final output of the MIMD is given by the difference between the outputs of the two subunits. Each leg or subunit of this MIMD is described by a pair of coupled ordinary differential equations, that is

$$\dot{e}_{\xi,1} = L_p(t) - k_1 e_{\xi,1},$$

$$\dot{e}_{\xi,2} = L_q(t) - k_2 e_{\xi,2}\left(1 + k_{12}f(e_{\xi,1})\right), \tag{5.10}$$

where $\xi = l, r$ depending on which leg (left or right) is being described; if $\xi = r$, $p = 1, q = 2$, otherwise $p = 2, q = 1$. Note that the MIMD is a delay-and-compare scheme in which the comparison operation is implemented via recurrent shunting inhibition. Some characteristics of the MIMD response will be considered next.

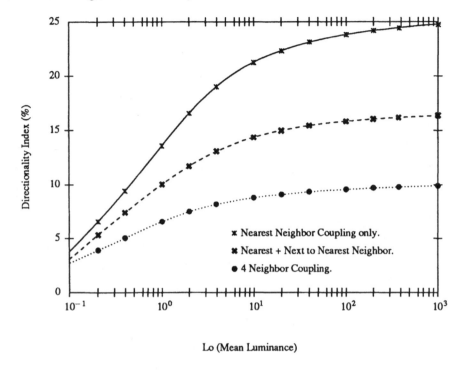

Figure 5.4   Percent Directionality Index as a function of mean luminance for a
bright bar of size = 1 comp, moving with a speed = 0.2 comp/sec, for networks
with nearest, nearest plus next to nearest and 4 nearest neighbor coupling.

5.4.3 Response Characteristics of MIMDs

The MIMD, due to its structural symmetry, does not respond to a uniform field of
illumination and responds to leftward and rightward motion with the same strength and
opposite sign. The response of the MIMD to an edge and a bar moving from left to
right, and vice versa, is illustrated in Fig. 5.8 on page 67.

Response to Drifting Gratings.   The response of the MIMD to drifting gratings
depends greatly upon the grating spatial and temporal frequencies and upon its contrast.
However, after the transient dies out the response to a drifting sine-wave grating usually
oscillates around an average response. If only second order nonlinearities are considered,
then the response of each leg of the MIMD, $e_{\xi,1}$ and $e_{\xi,2}$, to a stimulus $L_0 + c\ell(t)$, where
$0 \leq \|\ell(t)\| \leq L_0$ and $0 < c \leq 1$, can be written as:

$$e_{\xi,1} = x_0 + cx_{\xi,1}, \quad \text{and} \quad e_{\xi,2} \simeq y_0 + cy_{\xi,1} + c^2 y_{\xi,2}.$$

Substituting for $e_{\xi,1}$ and $e_{\xi,2}$ into (5.10) and collecting terms of like power of c, after
expanding $f(e_{\xi,1})$ into a Taylor series about x_0, yields the following:

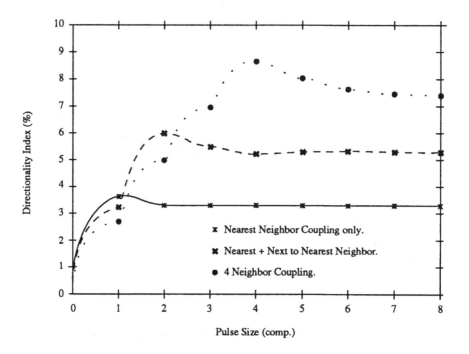

Figure 5.5   Percent Directionality Index as a function of size of a bright bar moving with a speed = 0.2 comp/sec, for networks with nearest, nearest plus next to nearest and 4 nearest neighbor coupling, with $L_0 = 0.1$ (arbitrary units).

$$
\begin{aligned}
x_0 &= L_0/k_1, & \dot{x}_{\xi,1} &= \ell_p(t) - k_1 x_{\xi,1}, \\
y_0 &= L_0/\alpha, & \dot{y}_{\xi,1} &= \ell_q(t) - k_2 k_{12} f'(x_0) y_0 x_{\xi,1} - \alpha y_1, \\
\alpha &= k_2(1 + k_{12} f(x_0)), & \dot{y}_{\xi,2} &= -(k_2 k_{12} f''(x_0) y_0/2) x_{\xi,1}^2 \\
& & & \quad - k_2 k_{12} f'(x_0) x_{\xi,1} y_{\xi,1} - \alpha y_{\xi,2}
\end{aligned}
\tag{5.11}
$$

Thus, an approximate solution to the MIMD response can be obtained by solving the above set of differential equations, that is

$$
m(t) \simeq c(y_{r,1} - y_{l,1}) + c^2(y_{r,2} - y_{l,2}).
\tag{5.12}
$$

Let Mr denote the MIMD mean response associated with second order nonlinearities. It can be easily shown, by solving Eq. (5.11), that Mr due to a moving sine-wave grating, $L(s,t) = L_0 + cL_0 \cos(2\pi\mu f_s s - 2\pi f_t t + \varphi)$, satisfies the relation

$$
Mr \propto \sin(\Delta\varphi)\sin(2\pi\mu f_s \Delta S),
\tag{5.13}
$$

where $\Delta\varphi = \tan^{-1}(2\pi f_t/\alpha) - \tan^{-1}(2\pi f_t/k_1)$. Equation (5.13) is similar to Eq. (5.7); thus, apart from a scale factor, the MIMD average response to drifting sine-wave gratings

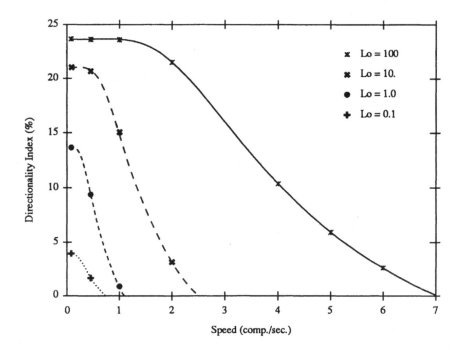

Figure 5.6 Percent Directionality Index as a function of speed of a moving, bright bar of size = 1 comp, for a unidirectional MLINN with nearest neighbor coupling: $\dot{e}_i = L_i(t) - e_i(1 + e_{i-1})$.

of low contrast is the same as that of the correlation model. This is due to the fact that both detectors are, up to second order, functionally equivalent.

The exact expression of Mr for a MIMD equipped with a lateral inhibition function, $f(e) = e$, is given by

$$Mr = \frac{k_2 k_{12} c^2 L_0^2 (k_1 - \alpha)\omega}{\alpha(k_1^2 + \omega^2)(\alpha^2 + \omega^2)} \sin(2\pi\mu f_s \Delta S), \qquad (5.14)$$

where c is the grating contrast, $\omega = 2\pi f_t$, f_t and f_s are the grating temporal and spatial frequencies respectively, L_0 is the mean luminance level, μ represents the direction of grating motion (1 for rightward motion and -1 for leftward motion), and $\alpha = k_2(k_1 + k_{12}L_0)/k_1$. The second order approximation is justified since for $f(e) = e$, odd order nonlinearities do not contribute to the mean response and a nonlinearity of order $2p$ adds a term in c^{2p}, which can be neglected for $p \geq 2$, especially for low contrast values. Equation (5.14) shows that the MIMD mean steady-state response is insensitive to contrast reversal and depends quadratically on contrast, c^2. Moreover, this mean

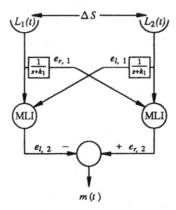

Figure 5.7 Multiplicative Inhibitory Motion Detector (MIMD). The outputs of the MLI boxes are: $\dot{e}_{r,2} = L_2(t) - k_2 e_{r,2}[1 + k_{12} f(e_{r,1})]$, and $\dot{e}_{l,2} = L_1(t) - k_2 e_{l,2}[1 + k_{12} f(e_{l,1})]$.

steady-state response peaks at the following spatial and temporal frequencies:

$$f_s = \frac{(2n+1)}{4\Delta S} \text{ (cycles/deg)}, \quad f_t = \sqrt{\frac{(k_1^2 + \alpha^2)\left(\sqrt{1 + 12k_1^2\alpha^2/(k_1^2 + \alpha^2)} - 1\right)}{24\pi^2}} \text{ (Hz)}$$

This means that the MIMD is tuned to certain spatiotemporal frequencies; by introducing spatial filtering at the receptor level, the MIMD can be tuned to a single spatiotemporal frequency.

5.5 DISCUSSION AND CONCLUSION

Models for visual motion information processing in biological and machine vision systems were divided into two categories: *intensity-based schemes* and *feature-matching schemes*. Some models from each category were discussed. Intensity-based models were further subdivided into *global* and *local* models. Other models have also been proposed that incorporate cooperative and/or competitive neural networks to infer motion (Bülthoff et al. 1989; Grossberg and Rudd 1989; Marshall 1990). These models use architectures of extensively interconnected elements and rely on several stages to process motion information.

Moreover, the role of *shunting lateral inhibitory neural networks* in mediating direction selectivity was investigated. In particular, the adaptation of the directional properties of these networks to mean luminance, and size and speed of moving objects was discussed. Recently a motion detector, known as the *multiplicative inhibitory motion detector (MIMD)*, has been proposed based on the directionality of shunting lateral inhibitory neural networks (Bouzerdoum and Pinter 1989a). This motion detector belongs to the family of intensity-based schemes and operates on the same basic structure as

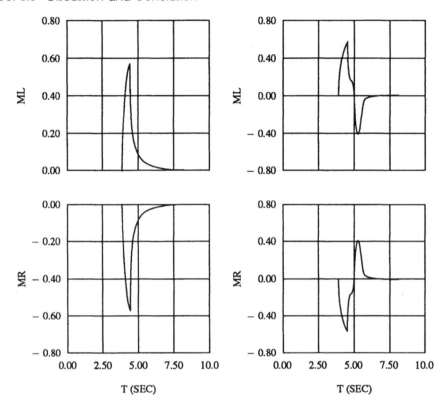

Figure 5.8   Response of MIMD to moving stimuli: positive edge (left) and positive bar (right). MR denotes the response $m(t)$ for rightward motion, and ML for leftward motion. $L_i(t) = L_0 + \frac{1}{2}L_0 u(t - t_0)$ for the edge, and $L_i(t) = L_0 + \frac{1}{2}L_0[u(t-t_0) - u(t-t_0-1)]$ for the bar, with $t_0 = 4$, or 4.5 sec. Mean luminance $L_0 = 10$, and speed = 2 comp/sec. Here, 1 compartmental unit (comp) = 1 ΔS (visual angle or length units).

the delay-and-compare schemes do, that is, a nonlinear asymmetric interaction between signals from two adjacent channels. However, the nonlinear interaction is of inhibitory nature originating from the biophysical mechanism of *shunting inhibition*.

The MIMD was shown to detect fairly well motion of objects like edges and bars. Furthermore, its mean steady-state response to a moving grating is insensitive to contrast reversal and it is equivalent to that of the Reichardt correlation model for gratings of low contrast. Examination of this response revealed that the MIMD can be tuned to a particular spatiotemporal frequency, hence it is a spatiotemporally oriented filter, which could perhaps be used in the implementation of energy models. Energy models may appear to be different from the delay-and-compare schemes, but it has been shown that the Reichardt correlation model is functionally equivalent to an energy model if its input stages include spatial receptive fields that are in quadrature (van Santen and Sperling

1984; Adelson and Bergen 1985). Finally, another important feature of the MIMD is its ability to adapt on changes of mean illumination levels; such adaptation was also found in motion sensitive interneurons in the third optic ganglion of the fly's brain (Dvorak et al. 1980; Borst and Egelhaaf 1987; de Ruyter van Steveninck et al. 1986). A preliminary report on the applicability of the MIMD to the activity of these neurons can be found in (Bouzerdoum and Pinter, 1990a); more details will be published soon.[3]

[3]This work is supported in part by National Science Foundation grants: BNS 8510188 & MIP 8822121.

6

Electronic Realization

In neural networks a large number of simple units perform operations that are locally simple but achieve globally complicated tasks. The inherent parallelism and fault tolerance of these networks, in addition to their unique processing capabilities, is readily adaptable to integrated circuit implementation. The following sections review some of the pioneering and representative work in *solid state* implementation of neural networks. The profusion of activity in implementation of neural networks, as evidenced by the number of papers presented, for example, at the meetings of *International Joint Conference on Neural Networks*, and *IEEE Neural Information Processing Systems: Natural and Synthetic* makes a thorough survey of the field impractical. A survey of recent activity may be found in (Card and Moore 1989). Furthermore, related topics such as optical implementations are not discussed.

6.1 ASSOCIATIVE MEMORY CHIPS

Researchers at AT&T Bell Laboratories were among the pioneers of solid state implementation of neural networks. Graf et al. (1986) fabricated an associative memory neural network chip which, given an incomplete vector finds the closest, in the Hamming distance sense, memory state. The memory states are local minima of the Liapunov (energy) function which arises from global stability analysis of the network. Each neuron performs a local computation by finding a weighted sum of the input lines and transforming

Matrix of Resistive Interconnections

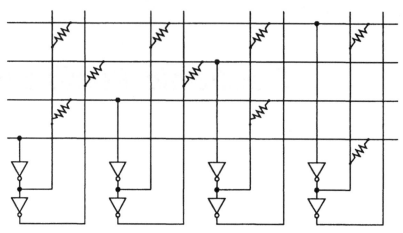

Array of Amplifier Units

Figure 6.1 A typical cross-bar network consisting of a matrix of resistive interconnections and an array of amplifier units ("neurons").

it via a sigmoidal, in this case a *sgn*, function,

$$v_i(t + 1) = \text{sgn}\left[\sum_j T_{ij} v_j(t)\right] \tag{6.1}$$

where

$$\text{sgn}(f) = \begin{cases} 1 & \text{if } f \geq 0 \\ 0 & \text{otherwise.} \end{cases}$$

It thus performs data processing (error correction), as well as data storage.

The chip was a direct translation of the cross-bar networks, also known as learning matrices, suggested by several authors[1] including Steinbuch (1963), Willshaw et al. (1969), Kohonen and Oja (1976), and Hopfield (1982, 1984), as shown in Fig. 6.1. The major innovation was the development of a process for fabrication of high density and high resistivity interconnection matrix—a resistivity of 100 $K\Omega$ and an impressive density of 4 resistors per square micron was achieved.

In an area of 5.7×5.7 millimeters, this chip accommodated 256 "neurons" with each neuron consisting of two inverting amplifiers to provide both positive and negative weights. There were a total of 25,000 transistors and 130,000 resistor sites, which provide trinary synaptic strengths $(0, \pm 1)$, and which could only be written once. This chip was thus a binary associative Programmable Read Only Memory (PROM).

The memory capacity of crossbar networks is between N/10 to N/4, with N being the total number of neurons. This chip can thus record and retrieve approximately 40

[1] For a review see Kohonen (1988).

memory states. The speed of retrieval depends on the input and the basins of attraction set up by local energy minima (the memory states), which in turn is determined by the matrix of connection strengths; the system may get trapped in spurious memory states or oscillate between two memory states. The average speed of recall was, however, a remarkable 700 nanoseconds.

As an associative memory circuit, the memory locations are determined by connection weights, the T_{ij} in Eq. (6.1), and thus the limited number of memory states and their lack of flexibility is a serious shortcoming. As a classifier such circuits need feature extractors prior to classification. The lack of specific purpose or application is common in almost all preliminary attempts in solid-state implementation of neural networks.

These problems were partially alleviated in a second design (Graf and de Vegver 1987) in which each connection strength was *controlled* by two memory cells. The connections could thus be programmed to assume any of the trinary values: excitatory, inhibitory or disconnected. This chip has 2916 such coupling elements between 54 fully interconnected neurons. With 2.5 micron design rules, it accommodated roughly 75,000 transistors on an area of 6.7×6.7 millimeters. With 10 stable states, each 40 bits long, programmed into the network, a recall capability of 20–700 nanoseconds was observed.

The increased flexibility of programmable connections was achieved at the cost of complexity and drastic reduction of capacity. Furthermore even though the association is a distributed operation and summation is performed via *analog* current summing, the chip has the peripheral circuitry of a standard digital static Random Access Memory (RAM).

Another innovative technology in implementation of neural networks was shown by Jet Propulsion Laboratory researchers (Thakoor 1987, Thakoor et al. 1986). In this implementation of a recurrent cross-bar network, a micro switch consisting of a switchable memory element and a high value resistor was placed at each cross section to provide binary strength interconnections. A planar microswitch which can be fabricated using .2-micron linewidths can contain as many as 4×10^8 nodes per squared centimeter. Delivery of a 1-microsecond pulse causes non-reversible switching; the circuit was thus an associative PROM. Two schemes, dilute coding and local inhibition, were proposed to increase the memory capacity and reduce recall error rates.

6.2 LEARNING CHIPS

An important feature of neural networks is their ability to modify their representation and long term memory storage of the external world, that is, to learn. Learning is usually achieved by changing the interconnection strengths via a *learning rule*. A design that accommodates variable connection weights is proposed by Mackie et al. (1988), where weights were stored as differences in voltages on two capacitors. A string of MOSFETs moved the charge between capacitors and thus varied the voltage differential. The chip itself did not learn; rather, it was designed to show the feasibility of incorporation of adaptive weights in a learning system.

A research group at Bell Communications Research has been among the first to concentrate on implementing learning networks. Alspector and Allen (1987) proposed the implementation of a *Boltzmann machine*. Boltzmann machines are recurrent networks in which each "neuron" thresholds a linear sum of its inputs but fires according to a probabilistic decision rule. A global minimum, as opposed to local minima of associative memory systems, and its corresponding weights determine a learned category. Boltzmann machines use noise to escape from local minima (incorrect representations), and reach global minima by varying a "temperature" parameter which is initially set high and then reduced as the system approaches global minima; hence the term *simulated annealing*. In this implementation thermal circuit noise is amplified to produce the annealing schedule. A 7×7 mm chip using 2 micron CMOS design rules was projected to require 250,000 transistors to produce a 40 neuron Boltzmann machine.

Several other groups are implementing another popular learning algorithm for multilayer networks, the error back propagation (Rumelhart et al. 1986), in VLSI.

6.3 SENSORY NEURAL NETWORKS

Despite the relative simplicity of the sensory neural networks compared to the central nervous system, there has been little published report on effort to capture the salient features of the sensory system. The pioneering work of Carver Mead's research group at California Institute of Technology is a notable exception.

A family of *silicon retinas*, fabricated in CMOS technology, has been described in Sivilotti et al. (1987), Mead and Mahowald (1988), and Mead (1989). The design philosophy has been to identify the main functionality of the anatomical units and to replicate them in solid-state circuits. To that end, the retinas consist of an innovative logarithmic photosensor layer proposed by Mead (1985), integrated with motion detecting (time derivative) units to mimic the presumed function of ganglion or amacrine cells, and spatial differentiators to represent horizontal and bipolar cells. Both the time and space derivative operations are performed by comparing the signal with its delayed and smoothed version.

The output of the chip shows many of the characteristics of bipolar cells in the vertebrate retinas; the temporal response is dependent on test flash diameter, intensity-response curves shift to higher order intensities at higher background illumination, and spatial edges are contrast enhanced.

6.4 DISCUSSION

Implementation of artificial neural networks has, within a very short period of time, resulted in innovations in technology, fabrication process, and analog and digital circuit design techniques, and has contributed to the understanding of biological systems as well as parallel computational techniques.

The models discussed above sample some of the preliminary, but representative, work in this area and as such have some shortcomings. The associative memory and learning circuits described above suffer from lack of purpose. They are, to some extent, solutions in search of a problem. Associative memories require extensive preprocessing, which limits their use, and are severely limited in application. For this reason they are used as special-purpose modules in an otherwise standard digital circuit. The learning circuits are intended to show the *feasibility* of learning on a chip, which is a necessary step for application to real-world problems. Furthermore, the relevance of these circuits to biological systems is limited.

The sensory neural networks discussed above, on the other hand, do try to mimic functional capabilities of biological systems by replicating some aspects of their neuroanatomy and electrophysiology. This approach has produced very interesting results but requires an advance knowledge of the role of each unit, and can suffer from a hasty preassignment of functionality to biological units. It also lacks in mathematical abstraction and in compatibility, that is, the peripheral processing does not, at least initially, lead to higher order processing. Many of these problems are being solved as the field matures (Mead 1989; Taylor 1990).

The next sections describe framework for electronic realization of the networks whose biological feasibility and relevance and self-sufficient processing capabilities were shown in the previous chapters.

6.5 DESIGN FRAMEWORK FOR ANALOG IMPLEMENTATION

The advantages, generality, and applicability of networks of multiplicative lateral inhibition as cited in Section 3.2 make their implementation very desirable, but digital implementation of these networks is very inefficient due to the need for:

- Analog to digital conversion, which can inherently suffer from aliasing when choosing sampling intervals and quantization error when assigning a number to the amplitude of the input signal.

- Multiplication and addition instructions.

- Registers, memories, address lines and other overhead circuitry.

- Implementation of iterative algorithms.

- Timing and other control circuitry.

- Long delays which may cause the system not to operate in real time.

Digital technology is a very mature technology with high noise immunity, great precision, and a very large library of available designs and functionalities that strive on the mathematically well-defined foundation of Boolean algebra. Had it not been for the distributed nature of the computations involved in a neural network that reduces the importance of the local computations, a digital implementation could have been preferable.

The choice between the two technologies is thus greatly dependent upon the nature of the problem; current summing can replace a digital adder only when the computation does not require the precision, and more importantly, the level of abstraction an adding *instruction* provides. For the present application analog implementation is ideal since the physical properties of the devices perform the desired computations. This conclusion is supported in Hopfield (1990) where digital and analog implementations are compared in terms of speed, area, and hardware precision.

6.6 DESIGN FRAMEWORK

Different functional modules that should generally be present in the implementation of shunting networks are as follows:

1. Input units, sensory or from antecedent layers.

2. Cell body with at least temporal and spatial integration capability.

3. Multiplicative interconnects for recurrent networks, preferably with variable, or programmable, connection strengths.

4. Multiplicative circuitry for *inputs* to implement nonrecurrent networks.

5. Summing circuitry.

6. Provision for both inhibitory and excitatory interaction.

7. Circuitry for nonlinear, preferably sigmoidal, transformation of feedback terms.

8. Interfacing units for correct polarity and fan-in/fan-out.

The very simple, extremely fast, all analog, and massively parallel circuit of Fig. 6.2 includes all of the above components with the exception of programmable connection weights, a design for which is presently under study. The following sections discuss the different components of this design framework.

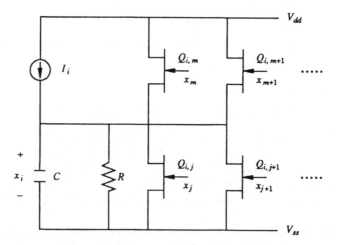

Figure 6.2 Design framework for implementation of one cell in a shunting network. A function of the voltage output of other cells is applied to the gate of transistors $Q_{i,j}$ which are drawn as n-channel JFET but can be any FET.

6.6.1 Input Sources

In the circuit of Fig. 6.2, *input units* are current sources that are either externally activated or arrive from antecedent cells or previous architectural layers of a multilayer neural network. For optoelectronic applications, which is the major proposed application area of the present work, the input current sources can be directly substituted by photodetectors since most photodetectors transfer optical input power directly to output current. Such detectors can be easily and compactly integrated monolithically with the FET circuitry of the neural network. An innovative design for the monolithic fabrication of a gallium arsenide photodetector which achieves almost twice the photosensitivity of a conventional detector with the same dimension and geometry, without any additional processing steps, has been described in Darling et al. (1988) and can be used in the integrated circuit implementation of the model. Silicon photodetectors whose photocurrent is proportional to the logarithm of light intensity and thus mimic the logarithmic response of many biological photoreceptors (Mead 1985; Delbrük and Mead 1989) are also well-suited for this circuit.

In the case where input sources are signals from previous architectural layers, since the output of a cell is a voltage, a voltage-to-current transformation is necessary. Many voltage-controlled current-sources have been described (Horowitz and Winfeld 1980), but this can simply be accomplished by placing a FET between the upper power supply rail and the appropriate cell. The *current* output of such a transistor is proportional to the square of the *voltage* applied to its gate-to-source terminal when operated in the current saturation region and thus provides the desired voltage-to-current transformation.

6.6.2 Cell Body and Temporal Characteristics

The cell body, which has also been called a "neuron" or a "processing unit" simply consists of a capacitor in parallel with a resistor, in direct analogy with the cell membrane capacitance and conductance of an iso-potential portion of a biological neuron. The capacitor in this simple circuit characterizes the temporal integration property of neurons. Spatial integration, or summation, is achieved by placing conductance inputs from other cells in parallel with the cell resistance.

For practical applications, if characterization of temporal properties is not of importance, the capacitor can be removed or in effect replaced with the parasitic capacitances present in any implementation. This allows for very high speed operation, possibly in gigahertz range for GaAs MESFETs. The parallel nature of the implementation means that the total speed of operation can be on the same order of magnitude as N times the speed of the operation of each cell, with N being the number of cells in the network. The speed of response is limited by the rate of convergence to a solution of the differential equations governing the behavior of the network. In Chapter 9 it will be shown that the implementation does indeed converge to a global solution, but the rate of convergence is not known. More severely, the major limiting factor in such networks is the readout of the output which will require multiplexing if the result is to be processed by a sequential machine. It is then best suited if higher-order processing is performed by a compatible system, as will be described later.

On the other hand, if accurate time dependencies need be studied, large capacitors may be necessary which are difficult to fabricate in integrated circuit form. In such cases the capacitor can be replaced by a circuit which compares, usually by subtraction, the present output with its delayed version and thus performs a crude differentiation operation. A circuit that achieves the latter task has been employed in (Sivilotti et al. 1987; and Mead and Mahowald 1988).

The output of such a neuron is a continuous valued, or analog, function which is analogous to the slow potential of biological neurons but not similar to the action potential stream (Kandel 1976) which is the primary method of communication in the nervous system. For modeling purposes such an output can be regarded as the *average firing rate* of a neuron whereby it is implicitly assumed that the absolute timing and amplitude of action potentials does not have processing significance. Gerstein et al. (1989) have studied correllograms to analyze the significance of the timing of action potential streams. Electronic circuit implementations which produce spike trains have been reported by Murray et al. (1989).

A more abstract view of the "cell potential" parameter is to regard it as the *number* of cells in a cell population which are active or "on" at a given time (Grossberg 1982). This interpretation thus assumes neurons of all-or-none character and has the advantage of describing the *statistics* of a *population* rather than the behavior of one neuron and thus uses stochastic analysis rather than deterministic techniques.

6.6.3 Shunting Recurrent Circuitry

The basis of the present design is the efficient implementation of multiplicative circuitry. This is achieved by using the voltage controlled conductance property of field-effect transistors when operated in the linear (ohmic) region. Figure 6.3 shows the current—voltage *(I—V)* characteristics of a typical FET.

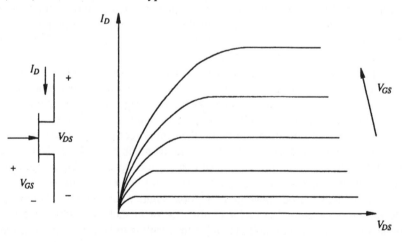

Figure 6.3   *I—V* Characteristics of a FET.

Operation of a FET in the linear (below saturation) region is described (Sze 1981) by

$$I_{DS} = K \left[(V_{GS} - V_{th})V_{DS} - \frac{1}{2}V_{DS}^2 \right] \quad \text{for} \quad V_{DS} < V_{GS} - V_{th}. \quad (6.2)$$

This region of operation is also known to circuit designers as the VVR: Variable-Voltage Resistance range.

In the current saturation region, the transistor operates as a voltage controlled current source, and is modeled by

$$I_{DS} = \frac{1}{2}K(V_{GS} - V_{th})^2, \quad \text{for} \quad V_{DS} > V_{GS} - V_{th}, \quad (6.3)$$

where I_{DS} and V_{DS} are drain-to-source current and voltages respectively, V_{GS} is gate-to-source voltage, V_{th} is the threshold voltage beyond which the transistor does not conduct, and K is a transconductance parameter dependent on doping density, active layer thickness, and gate width to length ratio.

As seen in Eq. (6.2) the input current is proportional to the product of V_{DS} and V_{GS}. It is this basic property that is used in the circuit of Fig. 6.2 to implement shunting networks because it allows a multiplication to be performed with *one* transistor only; a more compact implementation would be difficult to achieve. Detailed description of the behavior of different types of FETs in different regions of operation will be described in the next chapter.

The quadratic term V_{DS}^2 in Eq. (6.2) can be ignored if V_{DS} is small, or can be approximately cancelled by using other FETs of opposite type, such as p-type, in the upper rail of Fig. 6.2, but its presence is in fact advantageous since it implements a shunting "self-excitation" term which contributes to the desired on-center off-surround anatomy of the network.

The same basic principle is used in Petrie (1984) to implement multiplicative inhibitory networks. Designs for analog circuits which perform multiplication but are based on a differential input signal are explained in Mead (1985). Related circuits for the implementation of variable weights for possible inclusion in additive networks are described in Tsividis and Satyanarayana (1987). Circuits for cancellation of the quadratic nonlinearity and extension of the multiplication range can be found in Czarnul (1986).

By using Eqs. (6.2) and (6.3) in the circuit diagram of Fig. 6.2, it is a straight-forward matter to observe that in such a framework the general network equation,

$$\frac{dx_i}{dt} = \pm I_i \pm a_i x_i \pm x_i(K_i x_i) \pm x_i \sum_{j \neq i} K_{ij} x_j \pm \sum_j K_j x_j^2, \quad (6.4)$$

can be realized where x_i is a positive definite variable and K_i's and K_{ij}'s can be tailored during the fabrication process but are unalterable afterwards.

6.6.4 Shunting Non-Recurrent Circuitry

Equation (6.4) does not include non-recurrent terms of the form $x_i \sum I_j$. The circuit of Fig. 6.2 can be expanded to easily accommodate these feedforward terms. The basic design can be schematically shown as in Fig. 6.4, where for the sake of consistency inputs are taken to be currents.

In the circuit of Fig. 6.4, the resistor R_{i1} is unnecessary if the circuit is included as part of the framework of Fig. 6.2, but resistor R_i should be included for current to voltage conversion. The transistor Q_i is assumed to be an n-channel enhancement mode FET. A depletion mode transistor can be used for Q_i only if a level shifting operation is performed prior to the application of the input to the gate of Q_i.

Such a circuit implements the network equation

$$C_i \frac{dx_i}{dt} = \frac{V_{DD}}{R_{i1}} - a_i x_i + K_i R_i x_i (I_i) + \frac{K_i}{2} x_i^2 - K_i R_i x_i \left(\sum_{k \neq i} I_k \right) \qquad (6.5)$$

Two problems are immediately evident in the design of the circuit in Fig. 6.4. First, the self excitatory current I_i, which is shown as a current sink, is not easily implemented if it is the output of a current source, for example a photodetector. Secondly, the term $R_i \sum_k I_k$ can exceed the threshold voltage of the transistor and turn it off.

A design that can accommodate excitatory non-recurrent connections, and thus solve the first problem is shown in Fig. 6.5 where an analog inverter is used to provide correct polarity and a depletion-mode transistor is chosen to simplify the design. This circuit is described by the network equation

$$C_i \frac{dx_i}{dt} = \frac{V_{DD}}{R_{i1}} - a_i x_i + K_i R_i x_i (I_i) + \frac{K_i}{2} x_i^2. \qquad (6.6)$$

The design of an analog inverter, however, is not trivial and since the inhibitory connections automatically provide a self-excitatory x_i^2 term as shown in Eq. (6.5), a circuit with only inhibitory terms should suffice for most applications.

The problem of exceeding the threshold voltage of the multiplying transistor Q_i can be solved either by including a saturating nonlinearity after the summing circuitry,

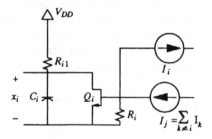

Figure 6.4   Schematic circuit diagram for the implementation of one cell in a non-recurrent network; I_i is total excitatory and I_j is total inhibitory input.

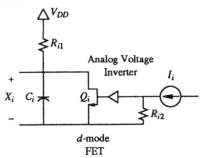

Figure 6.5   Circuit for the implementation of excitatory non-recurrent terms.

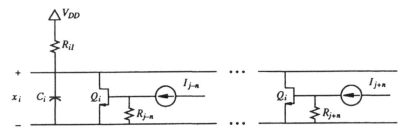

Figure 6.6   Circuit for the implementation of inhibitory nonrecurrent terms.

as will be shown later, or by using one transistor per input source to gate the currents, as shown in the circuit of Fig. 6.6. This circuit implements the network equation

$$C_i \frac{dx_i}{dt} = \frac{V_{DD}}{R_{i1}} - a_i x_i + \frac{K_i}{2} x_i^2 - x_i \left(\sum_{j \neq i} K_{ij} R_j I_j \right) . \tag{6.7}$$

which has the added advantage that since the connection strengths are defined by $K_{ij} R_j$, they can be selectively hard-wired, that is, the interconnection profile defined, by either varying the transconductance parameter or the resistor values.

A Fully Connected Feed Forward Network.   All of the designs that have been presented are adaptable to very high integration densities. This claim can be validated by examining the circuit of Fig. 6.7 where a crossbar network of resistive elements provides a fully connected interconnection matrix for an array of *input voltages, V_i*. As was described in the previous section such resistive networks have been fabricated with a density of 4 resistors per square micron (Graf et al. 1986), or even with switchable resistances with densities of 4×10^8 nodes per square centimeter (Thakoor et al. 1986). The circuit of Fig. 6.7 adds 2N resistors, N transistors, and possibly N capacitors to such resistive networks, with N being the number of neurons, without the need for any additional interconnect lines, and hence does not require much additional area.

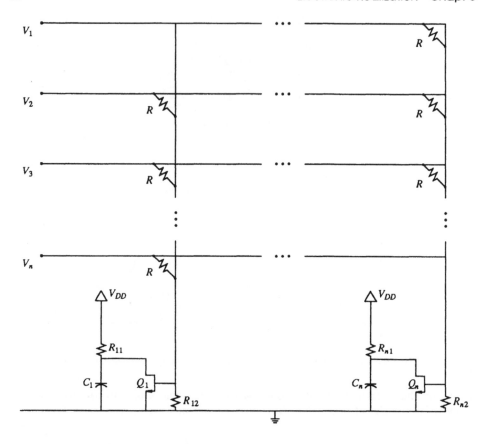

Figure 6.7 A fully connected feed forward shunting network.

The voltage presented to the gate of transistor Q_i of the circuit in Fig. 6.7 can be found by noting that

$$\sum_j \frac{V_j - V_{GS,i}}{R} = \frac{V_{GS,i}}{R_{i2}}, \tag{6.8}$$

where for simplicity it is assumed that all interconnect resistors have the same value. This leads to

$$V_{GS,i} = \frac{\sum_{j \neq i} V_j}{\frac{R}{R_{i2}} + N - 1}, \tag{6.9}$$

where N is the number of inputs connected to the cell i.

The total input presented to the cell i is thus divided by the denominator of Eq. (6.9) which is at least $(N-1)$, and can be further controlled by the choice of resistor ratio $\frac{R}{R_{i2}}$. Proper choice of this ratio allows a large dynamic range of the input to be processed by the cell.

If transistor Q_i is biased in the linear region, output of the cell i is described by

$$C_i \frac{dx_i}{dt} = \frac{V_{DD}}{R_{i1}} - a_i x_i + \frac{K_i}{2} x_i^2 - x_i b_i \left(\sum_{j \neq i} V_j \right), \qquad (6.10)$$

where

$$b_i = \frac{1}{\frac{R}{R_{i2}} + N - 1},$$

and

$$a_i = \frac{1}{R_{i1}} - K_i V_{th},$$

which is the desired shunting non-recurrent network equation. This discussion shows that with existing technologies this circuit is capable of implementing a compact and fully interconnected shunting nonrecurrent neural network.

6.6.5 Summing Circuitry and Excitatory and Inhibitory Connections

In the circuit of Fig. 6.2 summation is performed by using Kirchhoff's current law which arises from the conservation of charge principle and thus is valid with unlimited precision. The distinction between excitatory and inhibitory connections, or the polarity of the terms of the network equations, is then basically determined by the direction of current flow in that circuit.

As mentioned previously, and discussed in detail for nonrecurrent networks, implementation of the multiplicative excitatory connections requires an analog inverter, and for most applications is better provided by the quadratic self-excitatory terms which are due to transistor nonlinearities and are provided cost-free by the shunting inhibitory connections.

Furthermore, often the cell activity x_i should accept both positive and negative values, corresponding to the depolarization and hyperpolarization of biological neurons. When x_i changes sign the excitatory terms become inhibitory and vice versa. It is for this reason that the excitatory terms appear in the network equations and contribute to the stability of the network. In the implementations discussed, the cell activity is assumed to be positive definite and hence the implementation of excitatory terms is not as important.

Nonetheless it is important that the excitatory shunting terms can be implemented because in some applications the profile of the receptive field which is mainly described by the anatomy of the dendritic arborization, that is, the ratio of excitatory to inhibitory interactions, is crucial. Also, it is believed that in learning the excitatory connections are modified while the inhibitory ones remain constant, and hence provision for excitatory interaction should be made.

6.6.6 Sigmoidal Nonlinearity

The shape of the nonlinearity in the feedback loop, that is, the shape of $f(x_j)$ in a network such as, but not limited to,

$$\frac{dx_i}{dt} = -ax_i + (B - x_i)[I_i + f(x_i)] - x_i \left[J_i + \sum_{j \neq i} f(x_j) \right] \qquad (6.11)$$

has special significance. It has been shown (Grossberg 1973) that if $f(x_i)$ is sigmoidal, or *S* shaped, it can suppress noise, contrast enhance suprathreshold activities, and normalize total activity. A sigmoidal signal implies the existence of a threshold, a *quenching threshold* as termed by Grossberg (1973, 1983), below which activity is suppressed and above which activity is contrast enhanced. This transformation is identified with storage in Short Term Memory (STM). Grossberg analyzes the effect of the feedback nonlinearity on processing of the input data, based on deviation from linearity at each activity level.

Sigmoidal, or at least saturating, nonlinearities occur naturally in electronic devices. The *I—V* curves of a single FET as shown in Fig. 6.3 provide an excellent example of a saturating nonlinearity wherein not only the slope, but also the saturation point is easily controllable by the gate-to-source voltage. Figure 6.8 shows simple designs which use this natural property to perform the desired transformation.

Since at low values of x_i, $f(x_i)$ is linear, such a circuit will amplify both signal and noise for low values of signal. If a faster-than-linear transformation is needed for low values of signal, that is, a true *sigmoidal* curve, other circuits such as a differential amplifier pair which is described by a hyperbolic tangent, *tanh*, may be employed. The choice

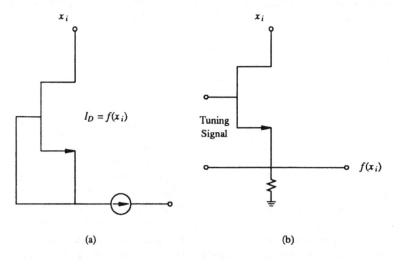

(a)   (b)

Figure 6.8   Simple circuits for implementation of a saturating nonlinear transformation; (a) shows the case where $f(x_i)$ is current, and (b) when it is a voltage value.

depends on the application and for the present purposes the simplest implementation seems the most appropriate.

6.6.7 Interface Circuitry

The large fan-in and fan-out associated with high connectivity neural networks and the importance of the polarity of the output necessitates a carefully chosen buffer stage.

If p-type or depletion mode devices are used in the basic cell diagram of Fig. 6.2, a level shifter is necessary for correct polarity. Implementation of excitatory connections requires an analog inverter.

Since the output of each cell connects to the gate of the other cells, which is a reverse biased diode, little current is drawn by fan-out connections and a design with only enhancement mode devices may be implemented even without a buffer stage.

Since each connection from other cells of the network provides a resistive path in parallel with the cell membrane resistance, fan-in considerations require careful determination of the amount of current flow per cell. For sensory networks which usually do not require very high connectivity, the designs suggested in this chapter are readily applicable. Details of these considerations for FET technology will be considered in Chapter 7.

6.7 THE CHOICE OF TECHNOLOGY

Since the basic cell of Fig. 6.2 is very similar to a standard logic gate inverter, but with transistors sized by gate width-to-length ratio to operate in the nonsaturated current region, the designs presented in this work are applicable to variety of field effect transistor technologies including NMOS, CMOS, and gallium arsenide (GaAs).

A circuit made of all depletion-mode devices such as GaAs MESFET buffered FET logic, can implement all the terms of Eq. (6.4) except shunting excitatory terms and requires a level shifter in the buffer stage. A design with all enhancement mode devices such as silicon NMOS can do the same but without a level shifter. With addition of p-channel devices, for example, silicon CMOS, all polarities and all terms of equations described in this chapter can be realized. For this case, as mentioned previously, a buffer stage is necessary for sufficient fanout.

6.7.1 Gallium Arsenide Framework

From the previous discussion it is evident that all the FET technologies can be used for the implementation of shunting neural networks. Silicon-based technologies are highly developed and have accumulated a very wide library of available designs and functionalities as well as highly efficient processing technologies. Gallium arsenide, on the other hand, has several long-term advantages relevant to the present application, including the following.

Gallium arsenide offers higher transconductance per unit gate width than silicon and can thus offer higher frequency/bandwidth operation, it is also capable of operating at high temperatures and is highly radiation tolerant. Since GaAs is a direct bandgap semiconductor, it is capable of efficiently emitting photons, as in an injection laser or a light emitting diode, as well as efficiently detecting photons in photodetectors such as photodiodes.   Both detectors and emitters, in addition to electronic circuitry, can be monolithically integrated on the same substrate, constituting integrated optoelectronic circuits (IOEC's).   Gallium arsenide also has larger piezoelectric, electro-optic, and electro-absorptive coefficients.

A capability unique to compound semiconductors is the possibility of "bandgap engineering" demonstrated by fabrication of heterojunctions, quantum wells, and super-lattices. Devices based on such structures offer extraordinary properties whose relevance to neural network applications may provide a fertile ground for future research. A design that uses this unique property by employing epitaxially grown aluminium gallium arsenide (AlGaAs) on a gallium arsenide substrate is presently under study.

On the negative side, gallium arsenide based devices suffer from several deficiencies which should be considered in the choice of the implementation medium. Silicon can form a native stable oxide with an atomically clean interface which is used in making Metal-Oxide-Semiconductor FETs (MOSFET). The MOSFET gate can then be biased with either polarity. Metal Semiconductor gate Field-Effect Transistors (MESFET's), on the other hand, are the device of choice for gallium arsenide technology since GaAs does not have a stable native oxide. The gate of a MESFET is a Schottky diode which should not be forward biased. It is for this reason that enhancement-mode devices are difficult to fabricate in GaAs technology, and hence the designs should primarily consist of depletion mode devices. This voltage swing limitation of enhancement mode devices is probably the most constraining for the implementation of neural networks; however, the circuit designs of this chapter have countered this problem by suggesting inverter and level-shifter circuitry.

Also, in gallium arsenide electrons are much more mobile than holes and thus p-channel devices require much greater area compared to n-channel devices of the same characteristics.   Hence GaAs circuits normally use only n-channel devices.   Even in n-channel devices the control of the threshold, and hence pinch-off, voltage poses a sensitive fabrication problem since it is defined by

$$V_p = \frac{qN_d a^2}{2\epsilon} \qquad (6.12)$$

where V_p is the pinch-off voltage, N_d is the majority carrier (donor) density, a is the active layer thickness, and q and ϵ are magnitude of electronic charge and semiconductor permittivity respectively.

It is seen from this relation that since high speed operation and large current drive requirements are met by high doping density of the *majority* carriers, the product $N_d a^2$ which determines the pinch-off voltage is very sensitive to active layer thickness which in turn requires strict control of both the growth and etching of this layer. This problem is somewhat alleviated by the fact that for analog neural network implementation, FET's should operate in the linear region and hence a large pinch-off voltage is desirable, which reduces the precision requirements of the active layer thickness. Interestingly, the speed of the operation is not sacrificed since the feedback mechanism of shunting nets, as described in Chapter 2, automatically limits the voltage swing and the transistor operates in a perturbation regime rather than having to turn on and off.

The last consideration in device technology trade-offs is that the high speed of GaAs circuits requires low loss conductors and hence interconnections require larger areas, which in turn increases parasitic nodal capacitances. Parasitic capacitances should be minimized by reducing both cross-sections of the interconnect lines and line to line capacitances; which are often contradictory goals.

The choice of technology, like most other engineering problems, requires a careful study of the trade-offs *as applied to a specific problem*. An important feature of the designs of this chapter is that they are applicable to *all* of the FET-based technologies, but based on the above mentioned trade-offs, silicon technologies are better suited for purely electrical applications of neural networks and offer the short-term advantages of available functionalities and streamlined fabrication facilities. For optoelectronic applications, GaAs and other compound semiconductors have very promising long term advantages. The final verdict can be made when the *unique capabilities* of each technology is employed and optimized. Chapter 7 analyzes applicability and trade-offs in choice of field-effect technology for neural net applications.

7

Implementation of
Shunting Networks
with FET Technologies

Robert B. Darling

7.1 ANALOG IMPLEMENTATIONS

Shunting recurrent neural networks may be phrased in terms of the general state equations of

$$\frac{dx_i}{dt} = -A_i x_i - x_i \sum_{j \neq i} f_{ij}^i(x_j) - I_i^i + x_i \sum_{k \neq i} f_{ik}^e(x_k) + I_i^e, \qquad (7.1)$$

each of which describes the evolution of the ith cell of the network (Grossberg 1988). The A_i term produces a self-relaxation of each state variable x_i and the I_i^i and I_i^e terms serve as inputs to the ith cell. Terms with the superscript i are inhibitory and those with the superscript e are excitatory to the time response of the state variable. The $f_{ij}^i(x_j)$ and $f_{ik}^e(x_k)$ are arbitrary, but monotonically increasing functions of the state variables x_j and x_k of neighboring cells and their presence in the summation indicates the existence of a unilateral inhibitory (excitatory) interconnection from the jth (kth) cell to the ith.

The process of implementing such a system of equations centers around finding a physical variable to which the x_i may be ascribed to in such a manner that the multiplicative operations and the nonlinear functions may be conveniently and compactly produced. Within digital electronic realizations, this representation of the state variables x_i becomes trivial, since the x_i are simply treated as variables in a computer program. However, such a realization is not an implementation in the true sense, because the state variables still do not have a one-to-one correspondence with any physical media. In short, such an implementation merely serves to re-cast the analytical mathematical abstraction into a digital hardware or software abstraction in which simulation can be easily accomplished. *Implementation* will here be carefully defined to mean a physical realization in which the physical variables of the implementation are interrelated in a manner that to first-order is described by the same set of differential equations as the proposed network.

Within electronic realizations, a physical implementation dictates that analog circuitry be used. Beyond this assertion, the method is without constraint and limited only by an engineer's creativity. Any of the traditional electronic variables of current, voltage, charge, magnetic flux, or field strength may be used to represent the state variables x_i. However, the accurate implementation of a large system of equations, such as the previously described shunting network system, can be best accomplished through the use of circuit techniques which capitalize upon the confinement and conservation of some quantity, such as charge. As usual, the use of circuit techniques allows the complex system to be decomposed into lumped elements plus an interconnection network which is far more easily dealt with than the three-dimensional interactions of spatially distributed devices. Inevitably, in an integrated circuit implementation, the most convenient choices are voltage and current.

The multiplication by x_i of both the inhibitory and the excitatory terms in these network equations is of central importance and necessary for the creation of a shunting network. Indeed, the implementation of this analog multiplication is by far the greatest concern in the design of an electronic realization of a shunting network. The design of the $f_{ij}^i(x_j)$ and $f_{ik}^e(x_k)$ functions is somewhat less crucial, so far as the function remains monotonically non-decreasing for stability considerations, and is, preferably, sigmoidal.

While numerous system advantages of the multiplicative terms, that is, of shunting networks in general, have been discussed in other chapters of this work, the present chapter will give a brief review of methods and constraints for achieving this through the current-voltage (I—V) characteristics of field-effect transistors (FETs). It should be noted that while a large number of other alternatives exist, such as bipolar transistors and various optical devices, FETs have several advantages. These include:

1. Implementation of the multiplication in a very direct manner.

2. Low power consumption with high integration densities.

3. Easy interfacing to a broad variety of other systems.

4. An already existing and well established technology for manufacture in the form of high density integrated circuits.

7.2 NONLINEAR DEVICE CHARACTERISTICS

In order to obtain some quantitative procedures for the design of shunting neural networks using integrated FET circuitry, it is first necessary to review the nonlinear device characteristics which allow FETs to implement the network state equations. The n-channel Si MOSFET and n-channel $GaAs$ MESFET will be examined, since other device variations will fall within these extremes.

7.2.1 N-Channel Si MOSFETs

An n-channel MOSFET is a representative device for neural network integration, being the staple for NMOS and CMOS circuitry. In depth derivations of the current-voltage (I—V) characteristics of MOSFETs may be found in standard references (Ong 1984; Tsividis 1987; Uyemura 1988), but only a summary of the applicable results is presented here. The MOSFET is constructed from a MOS capacitor which is used to control the mobile carrier density in a surface inversion channel of the semiconductor. The MOS capacitor consists of a heavily doped polysilicon gate electrode that is deposited over top of a thin layer of SiO_2. The back contact of the MOS capacitor is the semiconductor substrate which is uniformly doped with N_a acceptors for an n-channel MOSFET. The Fermi potential of the bulk semiconductor thus lies at

$$\phi_F = \frac{kT}{q} \ln \frac{N_a}{n_i} \tag{7.2}$$

above the intrinsic electrostatic potential of the substrate, where n_i is the intrinsic carrier concentration of the semiconductor and kT/q is the thermal voltage. Strong inversion of the semiconductor surface occurs when the energy bands are bent to an amount that just reverses the sign of the Fermi potential at the surface and corresponds to a potential drop of $2|\phi_F|$ across the resulting depletion region.

The thin oxide layer of SiO_2 produces an oxide capacitance of

$$C'_{ox} = \frac{\epsilon_{ox}}{x_{ox}}, \tag{7.3}$$

where ϵ_{ox} is the permittivity of the oxide and x_{ox} is the oxide thickness, typically 800 to 1200 Angstroms for modern MOSFET processes. The prime on the oxide capacitance denotes that it is per unit area.

The value of gate to substrate voltage required to level the energy bands is termed the flatband voltage,

$$V_{FB} = \Phi_G - \Phi_S - \frac{1}{C'_{ox}}(Q'_{ox} - Q'_{ss} - qD_I), \tag{7.4}$$

and is set by the difference of the work functions of the gate and of the semiconductor and by the presence of any fixed oxide charge or surface state charge. Additionally, an ion implant for threshold voltage adjustment is frequently used and this effectively adds an additional charge at the $Si - SiO_2$ interface of the amount qD_I, where D_I is the dose of the implant. This additional charge is usually included as an effective change in the flatband voltage as shown.

The bending of the energy bands to invert the surface of the semiconductor also produces a depletion region that extends inward from the surface of the semiconductor and which supports a depletion charge of

$$Q'_d = \sqrt{2q\epsilon_{Si}N_a(2|\phi_F|)}, \tag{7.5}$$

where ϵ_{Si} is the permittivity of the semiconductor. When an inversion layer is present this depletion region may be independently biased by an applied channel to body voltage V_{CB}. Both of these voltages are customarily referenced to the source of the device for which the depletion charge takes the form of

$$Q'_d = \sqrt{2q\epsilon_{Si}N_a(2|\phi_F| + V_{CS} - V_{BS})}. \tag{7.6}$$

The threshold voltage of the MOS capacitor is defined as that value of gate to source voltage required to place the surface channel just on the edge of strong inversion. This is equal to the gate to source voltage to first flatten the bands plus the amount to actually bend the bands by $2|\phi_F|$ plus an additional amount to support the depletion charge:

$$V_T(V_{CS}) = V_{FB} + 2|\phi_F| + \frac{1}{C'_{ox}}\sqrt{2q\epsilon_{Si}N_a(2|\phi_F| + V_{CS} - V_{BS})}. \tag{7.7}$$

The nominal threshold voltage of the MOS capacitor is that with zero channel to body bias,

$$V_{TO} = V_{FB} + 2|\phi_F| + \frac{1}{C'_{ox}}\sqrt{2q\epsilon_{Si}N_a(2|\phi_F|)}. \tag{7.8}$$

A body-bias parameter is defined as

$$\gamma = \frac{1}{C'_{ox}}\sqrt{2q\epsilon_{Si}N_a}, \tag{7.9}$$

from which

$$V_T(V_{CS}) = V_{TO} + \gamma\left(\sqrt{2|\phi_F| + V_{CS} - V_{BS}} - \sqrt{2|\phi_F|}\right). \tag{7.10}$$

When the gate voltage exceeds the threshold voltage, the additional increment supports the inversion charge of electrons which forms the conducting channel of the MOSFET. The inversion charge per unit area of the gate is

$$Q'_I = -qC'_{ox}(V_{GS} - V_T(V_{CS})). \tag{7.11}$$

Integrating this mobile carrier density over the area of the gate gives rise to the channel current I_D. The current flow is proportional to a process transconductance parameter, defined as

$$k' = \mu_n C'_{ox}, \tag{7.12}$$

where μ_n is the electron mobility. After integrating over the gate area, a device transconductance parameter of

$$k = \frac{W}{L}k' = \mu_n C'_{ox}\frac{W}{L}, \tag{7.13}$$

arises, where W and L are the width and length of the MOSFETs gate and constitute the principal layout parameters used in the design of FET circuits. While the oxide capacitance is relatively well defined, the electron mobility used in the above relations refers explicitly to the mobility of electrons existing in the surface inversion layer of the MOSFET which is subject to a variety of processing parameters and is also difficult to accurately measure. As such, it often becomes a fitting parameter for a given MOSFET fabrication process, although it rarely differs from half of the bulk mobility of Si by more than 30 percent.

Integrating along the length of the channel in what is termed the "exact" gradual channel analysis yields the integral for the drain (or channel) current of

$$I_D = k \int_0^{V_{DS}} [V_{GS} - V_T(V_{CS}) - V_{CS}] \, dV_{CS}. \tag{7.14}$$

Substituting in the position dependent threshold voltage gives

$$I_D = k \left\{ (V_{GS} - V_{TO} + \gamma(2|\phi_F|)^{1/2})V_{DS} - \tfrac{1}{2}V_{DS}^2 \right.$$
$$\left. - \tfrac{2}{3}\gamma \left[(2|\phi_F| + V_{DS} - V_{BS})^{3/2} - (2|\phi_F| - V_{BS})^{3/2} \right] \right\}. \tag{7.15}$$

The above equation provides a quite accurate representation for the channel current of a MOSFET under conditions of strong inversion in the channel. When the drain to source bias is sufficiently increased, the inversion charge on the drain side end of the channel will be reduced below the point of strong inversion. This results in saturation of the channel current and the above expression is no longer valid. The customary means for extending the analysis into the saturated region has been to take the point where the nonsaturated analysis gives a maximum channel current and define this point as the boundary between the saturated and nonsaturated regions of operation. In the saturated region, the $I_D - V_{DS}$ characteristics are then extrapolated from the boundary with the nonsaturated region. This is expressed mathematically by defining a value for V_{DS} that places the FET on the edge of saturation,

$$\left. \frac{\partial I_D}{\partial V_{DS}} \right|_{V_{DS}=V_{DS,sat}} = 0, \tag{7.16}$$

and this gives an extrapolated drain current of

$$I_{D,sat} = I_{D,nonsat}(V_{DS} = V_{DS,sat}) \tag{7.17}$$

for the saturation region.

For the above nonsaturated MOSFET channel current, this procedure leads to a drain to source saturation voltage of

$$V_{DS,sat} = (V_{GS} - V_{TO} + \gamma(2|\phi_F|)^{1/2})$$
$$- \tfrac{1}{2}\gamma^2 \left[\sqrt{1 + \tfrac{4}{\gamma^2}(V_{GS} - V_{TO} + \gamma(2|\phi_F|)^{1/2} + 2|\phi_F| - V_{BS}} - 1 \right]. \tag{7.18}$$

This turns out to be quite important in the design of shunting networks since it gives the signal swing limits for each region of operation.

The saturated drain current corresponding to the above $V_{DS,sat}$ is too complicated to merit citing, however it can be easily obtained by substitution of the above $V_{DS,sat}$ value. This representation of the saturated region of operation produces a saturated drain current which is independent of V_{DS}. In practice, a nonzero output conductance will arise from a variety of physical effects, the most dominant for MOSFETs being channel length modulation. At the edge of saturation, the drain edge of the channel is just at the point of strong inversion and the gate to channel potential difference is set solely by the depletion charge Q'_d. Using the depletion approximation to model the saturation region beyond this point results in the inversion region of the channel being reduced to an effective length of $L - \Delta L$ where the reduction in channel length is

$$\Delta L \approx \sqrt{\frac{2\epsilon_{Si}}{qN_a}(V_{DS} - V_{DS,sat})}. \tag{7.19}$$

The decrease in the effective length of the channel with increasing $(V_{DS} - V_{DS,sat})$ causes an increase in the channel current,

$$I_D = \frac{I_{D,sat}}{\left(1 - \frac{\Delta L}{L}\right)} \approx I_{D,sat}(1 + \lambda V_{DS}). \tag{7.20}$$

This is the principal contribution to the output conductance of a saturated MOSFET.

The above description is generally quite accurate for both the saturated and nonsaturated regions of MOSFET operation; however, the equations are rather unwieldy for circuit analysis. A simpler set is the model of Shichman and Hodges (1968):

$$I_D \begin{cases} \beta[(V_{GS} - V_T(0))V_{DS} - \frac{1}{2}V_{DS}^2] & \text{for } 0 < V_{DS} < V_{DS,sat}, \text{ (nonsaturated);} \\ \frac{1}{2}\beta(V_{GS} - V_T(0))^2(1 + \lambda V_{DS}) & \text{for } V_{DS,sat} < V_{DS}, \text{ (saturated),} \end{cases}$$

where the threshold voltage at the source side of the channel is

$$V_T(0) = V_{TO} + \gamma(\sqrt{2|\phi_F| - V_{BS}} - \sqrt{2|\phi_F|}), \tag{7.21}$$

and the drain to source saturation voltage is

$$V_{DS,sat} = V_{GS} - V_T(0). \tag{7.22}$$

While the Shichman-Hodges model is quite tractable for circuit analysis, it unfortunately differs significantly from the result of the more exact analysis. In particular, the point of saturation is significantly overestimated, both in drain current and in drain-to-source voltage. There are a variety of ways of resolving this. The simplest method is to scale the transconductance parameter to produce the same saturated drain current for both models (Uyemura 1988):

$$\beta = k\frac{I_{D,sat}(\text{exact analysis})}{I_{D,sat}(\text{circuit equations})}. \tag{7.23}$$

This is a viable approach for digital circuits which predominantly operate in either the saturated or cutoff mode. However, the output conductance for small values of V_{DS} is underestimated and $V_{DS,sat}$ is overestimated by this approach.

Another method that is more accurate is to linearize the depletion charge's dependence upon the channel voltage which is the source of the mathematical complications:

$$Q'_d \approx C'_{ox}\gamma\sqrt{2|\phi_F| - V_{BS}} + C'_{ox}\delta V_{CS}, \qquad (7.24)$$

where

$$\delta = \frac{\partial}{\partial V_{CS}}\left(\frac{Q'_d}{C'_{ox}}\right). \qquad (7.25)$$

This was first done by Merckel, Borel, and Cupcea (1972) and is also discussed in detail by Tsividis (1987). Using this, the drain current integral becomes

$$I_D = k\int_0^{V_{DS}}[V_{GS} - V_T(0) - (1 + \delta)V_{CS}]\,dV_{CS}, \qquad (7.26)$$

which produces an overall MOSFET I—V model of

$$I_D = \begin{cases} k[(V_{GS} - V_T(0))V_{DS} - \frac{1}{2}(1 + \delta)V_{DS}^2] & \text{for } 0 < V_{DS} < V_{DS,sat}, \text{ (nonsaturated)}; \\ \frac{1}{2}k\frac{1}{(1+\delta)}(V_{GS} - V_T(0))^2(1 + \lambda V_{DS}) & \text{for } V_{DS,sat} < V_{DS}, \text{ (saturated)}, \end{cases}$$

where

$$V_{DS,sat} = \frac{V_{GS} - V_T(0)}{1 + \delta}. \qquad (7.27)$$

This model provides good accuracy both in the saturation region as well as in the nonsaturated region for small drain to source voltages. The are a large number of methods for determining the parameter δ; but these are beyond the scope of this discussion and are summarized well by Tsividis (1987). This model is generally the most applicable for the analysis of shunting networks.

It should be noted that a number of neural network implementations have made effective use of the MOSFET I—V characteristics in the subthreshold region where $V_{GS} < V_T(0)$, (e.g., Lazzaro et al. 1989). For completeness, the applicable MOSFET characteristics will be cited here. The basic subthreshold model assumes that the channel is weakly inverted along its entire length and that conduction is due only to diffusion of the carriers. The most convenient results are those derived by Van Overstraeten et al. (1975) which assume an average energy band bending of $\frac{3}{2}|\phi_F|$ and expand the characteristics about this point. The result is a subthreshold drain current of

$$I_D = k\left(\frac{kT}{q}\right)^2 \xi e^{-q|\phi_F|/2kT} e^{q(V_{GS} - V_w)/(1+\xi)kT}\left(1 - e^{-qV_{DS}/kT}\right), \qquad (7.28)$$

where the gate to source voltage required to produce three quarters of the strong inversion band bending at the source side of the channel is

$$V_w = V_{FB} + \tfrac{3}{2}|\phi_F| + \gamma\sqrt{\tfrac{3}{2}|\phi_F| - V_{BS}} \qquad (7.29)$$

and

$$\xi = \frac{\gamma}{2\sqrt{\tfrac{3}{2}|\phi_F| - V_{BS}}}. \qquad (7.30)$$

This subthreshold model is valid for both saturated and nonsaturated channel currents and it should be noted that $V_{DS,sat}$ is approximately three times the thermal voltage or $3\frac{kT}{q}$; hence, most circuits employing subthreshold conducting MOSFETs operate with saturated channel currents.

7.2.2 N-Channel GaAs MESFETs

Another representative device for the implementation of shunting networks is the $GaAs$ MESFET. This device provides a contrast in characteristics as compared to the Si MOS-FET, since in general $GaAs$ MESFETs allow for much higher operating frequencies (10s of GHz versus 100s of MHz) as a result of higher transconductance and lower input capacitance, but this is accomplished at the expense of greatly increased current densities and consequently greater power dissipation. The maximum integration densities of $GaAs$ MESFETs are currently only about one tenth to one eighth of that of Si MOSFETs. Thus, these two devices provide a wide difference in characteristics when applied to the implementation of LSI-level circuitry.

In contrast to a MOSFET which modulates channel conduction by means of changing the carrier density in a surface inversion channel, a MESFET or JFET modulates the flow of current in a semiconductor channel by means of changing the undepleted cross-sectional area of the channel. Parallel to the previous MOSFET description, only an n-channel JFET or MESFET will be considered. The situation of an n-type channel layer of thickness a and uniformly doped with donors at a density of N_d provides the simplest model. For the case of a MESFET, the gate electrode forms a rectifying or Schottky contact to the semiconductor which then possesses a built-in voltage of

$$V_{bi} = \phi_{Bn} + \frac{kT}{q}\ln\left(\frac{N_d}{N_C}\right), \qquad (7.31)$$

where ϕ_{Bn} is the Schottky barrier height and N_C is the effective density of states in the conduction band. For the case of a JFET, the gate consists of a pn-junction whose built-in voltage is

$$V_{bi} = \frac{kT}{q}\ln\left(\frac{N_d N_a}{n_i^2}\right), \qquad (7.32)$$

where n_i is the intrinsic carrier concentration and N_a is the acceptor doping density of the gate. The I—V characteristics that will be presented below apply equally well to both the JFET and the MESFET, with the only difference coming from the different built-in voltages of the two types of gates.

For either the JFET or the MESFET, the conducting channel cross-section is determined by the undepleted thickness of the channel layer. The depletion depth h into the channel layer is controlled by the gate bias and the potential of the channel at the other side of the junction, both usually referenced to the source electrode,

$$V_{bi} - V_{GS} + V_{CS} = \frac{qN_d h^2}{2\epsilon_s}. \tag{7.33}$$

The above expression is termed the gating function for the channel. The application of sufficient reverse bias to the gate will extend the depletion region completely through the channel layer, and the necessary amount of band bending in the semiconductor for this to occur is termed the internal pinchoff voltage,

$$V_P = \frac{qN_d a^2}{2\epsilon_s}. \tag{7.34}$$

At this point, the depletion depth is equal to the layer thickness, $h = a$. Referenced to the source electrode, the value of gate to source potential required to put the channel just on the edge of conduction is termed the threshold voltage,

$$V_T = V_{bi} - V_P. \tag{7.35}$$

For the more common depletion-mode FETs, the magnitude of the pinchoff voltage is greater than the built-in voltage so that the threshold voltage is negative.

It is common practice to defined normalized depletion depths at the source end, interior, and drain end of the channel, respectively, as

$$u_1 = \sqrt{\frac{V_{bi} - V_{GS}}{V_P}}, \tag{7.36}$$

$$u = \sqrt{\frac{V_{bi} - V_{GS} + V_{CS}}{V_P}}, \tag{7.37}$$

$$u_2 = \sqrt{\frac{V_{bi} - V_{GS} + V_{DS}}{V_P}}. \tag{7.38}$$

For a gate width of W, the cross-sectional area of the channel that may conduct current is thus $Wa(1 - u)$. The channel current at any point along the channel is then given by

$$I_D = q\mu_n N_d W a(1 - u)\frac{dV_{CS}}{dx}. \tag{7.39}$$

The short gate lengths L and higher mobility of GaAs MESFETs or JFETs normally place the carrier transport of the channel into the hot electron range for which carrier velocity saturation effects must be included. This can most easily be introduced through the use of a field dependent mobility of the form

$$\mu_n = \frac{\mu_{no}}{1 + \dfrac{\mu_{no}}{v_{sat}}\dfrac{dV_{CS}}{dx}}, \tag{7.40}$$

where μ_{no} is the low-field mobility of the semiconductor and v_{sat} is the scattering-limited electron velocity.

Substituting this along with the gating function into the equation for channel current and integrating over the length of the channel produces the commonly-quoted model of Lehovec and Zuleeg (1970) which is valid for nonsaturated channel currents:

$$I_D = GV_P \frac{(u_2^2 - u_1^2) - \frac{2}{3}(u_2^3 - u_1^3)}{1 + z(u_2^2 - u_1^2)}. \qquad (7.41)$$

The undepleted channel conductance is

$$G = q\mu_{no}N_d a \frac{W}{L}, \qquad (7.42)$$

and a dimensionless parameter characterizing the influence of velocity saturation is

$$z = \frac{\mu_{no}V_P}{v_{sat}L}. \qquad (7.43)$$

This result for nonsaturated channel currents may be re-expressed in terms of the terminal voltages as

$$I_D = \frac{G\left[V_{DS} - \frac{2}{3}V_P^{-1/2}\left\{(V_{bi} - V_{GS} + V_{DS})^{3/2} - (V_{bi} - V_{GS})^{3/2}\right\}\right]}{1 + z\dfrac{V_{DS}}{V_P}}. \qquad (7.44)$$

Within this model, saturation of the channel current occurs when all of the electrons at the drain edge of the channel are moving at their scattering-limited velocity; therefore, the saturation point is fixed at

$$I_{D,sat} = qN_d W v_{sat} a(1 - u_{2s}). \qquad (7.45)$$

Additionally

$$V_{DS,sat} = V_{GS} - V_{bi} + V_P u_{2s}^2, \qquad (7.46)$$

where u_{2s} is the maximum normalized depletion depth at the drain edge of the channel required to sustain the saturated value of drain current. In other words, the channel closes down to a minimum opening of $a(1 - u_{2s})$. The value of u_{2s} is a function of the gate to source voltage and is determined through the transcendental equation

$$u_{2s}^3 + 3u_{2s}\left(\frac{1}{z} - u_1^2\right) + 2u_1^3 - \frac{3}{z} = 0. \qquad (7.47)$$

This model provides excellent agreement with the measured characteristics of uniformly doped GaAs MESFETs; however, the above equations are obviously rather awkward to use. Curtice (1980) has demonstrated that the simple analytical expression

$$I_D = \beta(V_{GS} - V_T)^2 \tanh(\eta V_{DS})(1 + \lambda V_{DS}) \qquad (7.48)$$

can provide a good fit to the measured device characteristics with suitable choices for the parameters of β, η, and λ. The above expression allows an easy interpretation of the parameters, that is,

$$I_{D,sat} = \beta(V_{GS} - V_T)^2, \tag{7.49}$$

$$\eta = \left.\frac{\partial I_D}{\partial V_{DS}}\right|_{V_{DS}\to 0;\ \text{(nonsaturation)}}, \tag{7.50}$$

$$\lambda = \left.\frac{\partial I_D}{\partial V_{DS}}\right|_{V_{DS}\to\infty;\ \text{(saturation)}}. \tag{7.51}$$

The parameter λ gives the output conductance of the FET in the saturated region for which most existing analytical models predict most poorly and as a result, λ almost always becomes an empirically determined parameter. However, the analytical models of FETs provide good accuracy in the nonsaturated region and as a result, the parameters β and η may be determined from them.

While the details are beyond the present scope of interest, the results are that β and η may be closely approximated as

$$\beta = \frac{G}{V_P}\frac{m(z)}{z}, \tag{7.52}$$

and

$$\eta = \frac{z}{m(z)V_P}\frac{1-\sqrt{1-x}}{x^2}, \tag{7.53}$$

where

$$m(z) = \tfrac{1}{2}(1 - e^{-z/2}) \tag{7.54}$$

and the normalized gate bias is

$$x = \frac{V_{GS} - V_T}{V_P}. \tag{7.55}$$

It may be noted that $x = 1 - u_1^2$. Within this approximation, the saturation voltage takes on a much simpler form,

$$V_{DS,sat} = \frac{4m(z)}{z}(V_{GS} - V_T). \tag{7.56}$$

This particular formulation has the advantage of reducing to the more standard case of no velocity saturation in the limit as $z \to 0$. This produces the result of

$$V_{DS,sat} = V_{GS} - V_T \tag{7.57}$$

using

$$\lim_{z\to 0}\frac{m(z)}{z} = \tfrac{1}{4}. \tag{7.58}$$

Once again, the above expressions for $I_{D,sat}$ and $V_{DS,sat}$ are important for determining the limits to signal swing in the saturated and nonsaturated regions of operation.

7.3 IMPLEMENTATION OF NONSATURATING FETS

The multiplication of the inhibitory and excitatory interconnection terms by the state variable of the cell x_i is the primary identifying feature of a shunting network. The importance of this multiplicative factor is discussed in other chapters of this book and will not be elaborated upon here. This section will focus instead upon the implementation of this multiplicative term using the current-voltage characteristics of both MOSFETs and MESFETs.

A current-voltage representation of the shunting network equations may be most directly obtained through a parallel connection of a set of controlled conductances for each ith cell whose common node voltage V_i is assigned as the state variable x_i. Any fixed conductance between the output node and ground provides the self-relaxation term $-A_i x_i$, and any capacitance contributes to the time derivative term $\frac{dx_i}{dt}$. Controlled conductances may be connected from the output node to ground to form inhibitory interconnections, or between the output node and a positive power supply rail V_{DD} to form excitatory interconnections. Current sources may inject charge into the node as an excitatory signal source, or extract charge from the node as an inhibitory signal source. The node equation for such a connection produces the network equation for the (ith) cell,

$$C_i \frac{dV_i}{dt} = -G_i V_i - V_i \sum_{j \neq i} G_{ij}^i(V_j) - I_i^i + (V_{DD} - V_i) \sum_{k \neq i} G_{ik}^e(V_k) + I_i^e, \qquad (7.59)$$

where G_{ij}^i and G_{ik}^e are the controlled conductances in the ith cell which are controlled by the state variables of the jth and kth cells. As before, the sum over j is over the inhibitory interconnections and the sum over k is over the excitatory ones.

The most readily available form of the controlled conductances are field-effect transistors, and most fortunately, these are also capable of high density monolithic integration via well-established fabrication processes. The channel connecting the drain to the source directly forms a conductance which can be modulated by the applied gate bias. Thus, the state variable is in effect the drain-to-source voltage of the inhibitory interconnection transistor,

$$V_i = V_{DSi}^i = V_{DD} - V_{DSi}^e. \qquad (7.60)$$

When an FET channel is substituted for each of the controlled conductances that make up the network interconnections, the network equations become

$$C_i \frac{dV_{DSi}^i}{dt} = -G_i V_{DSi}^i - \sum_{j \neq i} I_{Dij}^i(V_{GSi}^i, V_{DSi}^i) - I_i^i + \sum_{k \neq i} I_{Dik}^e(V_{GSi}^e, V_{DSi}^e) + I_i^e. \qquad (7.61)$$

However, the most common electronic uses of a field-effect transistor, that is, as a switch or an amplifier, place it into the saturated region of current flow where the channel current is a weak function of the drain-to-source voltage. This is not appropriate for the implementation of a shunting network because the multiplicative term in the network equations is then lost. Instead, the FET must be operated in the nonsaturated or ohmic region. This is often termed the *voltage-variable resistor* or VVR region and appropriately describes the function of the FET in implementing a cell-to-cell interconnection. Even so, saturating FETs are not without their uses in implementing shunting networks. Their characteristics more nearly resemble an ideal current source and can effectively function as injectors or extractors of current to implement the I_i^i or I_i^e cell input terms. Thus, if a given interconnection FET is allowed to saturate, it will change from implementing a multiplicative interconnection to producing a source term that depends upon the node voltage of some other cell.

Because the required dynamic range of most neural circuits extends over several decades, it is necessary in the design of such networks to maximize the voltage range over which the FET remains in the nonsaturated region of operation. For the general case, the channel current of any FET may be expressed as

$$I_D = V_{DS} G(V_{GS}, V_{DS}) \tag{7.62}$$

by simply factoring out V_{DS} from the previously cited I—V characteristics of the device. Saturation of the channel current with increasing V_{DS} leads to

$$\frac{\partial}{\partial V_{DS}} G(V_{GS}, V_{DS}) < 0. \tag{7.63}$$

The conductance function $G(V_{GS}, V_{DS})$ represents a perfect voltage-variable resistance when it is only a function of the input V_{GS},

$$I_D = V_{DS} G(V_{GS}). \tag{7.64}$$

In the limit of vanishingly small V_{DS}, all FETs exhibit I—V characteristics that conform to this description near the origin of the I_D—V_{DS} axes. Here, the conductance function may be regarded as simply the Taylor-series coefficient of

$$G(V_{GS}) = \left. \frac{\partial I_D}{\partial V_{DS}} \right|_{V_{DS}=0}. \tag{7.65}$$

However, as V_{DS} becomes more than infinitesimally small, the channel current of the FET will begin to saturate, reducing the magnitude of the conductance function and making the FET channel behave more like an ideal current source. The goal in device design is therefore to make the onset of saturation occur at as large a V_{DS} bias as is required to accomodate the signal swing of the output node.

Maximization of the nonsaturation region amounts to maximizing the drain-to-source saturation voltage $V_{DS,sat}$ for a particular value of channel current. The level of channel current itself can be easily adjusted through the width to length ratio of the gate. The $\frac{W}{L}$ factor thus provides arbitrary scaling of the I_D—V_{DS} characteristics along the current axis. Because $V_{DS,sat}$ decreases toward zero as $I_D \to 0$, it is desirable to not only scale the I—V characteristics of the device along the I_D and V_{DS} axes, but also to change the shape of the characteristics near the origin to extend the nonsaturation region, particularly for the case of small channel currents.

The first case to be considered is that of the n-channel MOSFET. Dividing the nonsaturated I—V characteristics by V_{DS} yields a conductance function of

$$G(V_{GS}, V_{DS}) = k[V_{GS} - V_T(0) - \tfrac{1}{2}(1 + \delta)V_{DS}]. \tag{7.66}$$

The saturation voltage was previously given as

$$V_{DS,sat} = \frac{V_{GS} - V_T(0)}{1 + \delta}. \tag{7.67}$$

From these it is apparent that the V_{DS} dependence of the conductance function can never be completely eliminated; however, it can be reduced by making the parameter δ as small as the processing technology makes practical. Similarly, a small value of δ helps to achieve the maximum $V_{DS,sat}$ for a given gate bias.

Directly calculating δ as the partial derivative of the depletion charge results in

$$\delta = \frac{1}{2C'_{ox}} \sqrt{\frac{2q\epsilon_{Si}N_a}{2|\phi_F| - V_{BS}}}. \tag{7.68}$$

Qualitatively, the nonsaturation region of a MOSFET is increased by using light substrate dopings N_a, thin gate oxides to increase C'_{ox}, and strong, negative body bias voltages V_{BS}. Realistic processing constraints will in practice limit the degree to which each of these may be applied.

Next, consider an n-channel MESFET. Dividing the nonsaturated I—V characteristics by V_{DS} results in

$$G(V_{GS}, V_{DS})$$
$$= \frac{G\left[1 - \tfrac{2}{3}V_P^{-1/2}V_{DS}^{-1}\left\{(V_{bi} - V_{GS} + V_{DS})^{3/2} - (V_{bi} - V_{GS})^{3/2}\right\}\right]}{1 + z\dfrac{V_{DS}}{V_P}}. \tag{7.69}$$

While accurate, this expression is awkward and it is easier to employ the previous model of Curtice (1980),

$$G(V_{GS}, V_{DS}) = \beta(V_{GS} - V_T)^2 \frac{\tanh(\eta V_{DS})}{V_{DS}}. \tag{7.70}$$

For a given gate bias,

$$V_{DS,sat} = \frac{4m(z)}{z}(V_{GS} - V_T) \tag{7.71}$$

and

$$I_{D,sat} = \frac{Gm(z)}{zV_P}(V_{GS} - V_T)^2. \tag{7.72}$$

As in the case of the MOSFET, the dependence of the conductance function on V_{DS} can never be completely eliminated, but it can be reduced by making the parameter η

smaller. This in turn is put into effect by making the parameter z as small as possible and using large pinchoff voltages V_P. Since

$$z = \frac{\mu_{no} V_P}{v_{sat} L},\qquad(7.73)$$

this requires that the MESFET have a long gate to reduce the effects of carrier velocity saturation. Qualitatively, carrier velocity saturation tends to make FET I-V characteristics appear more "bipolar-like" by reducing the size of the nonsaturation region. For the implementation of shunting networks, just the opposite is desired and carrier velocity saturation should be reduced as much as possible. Additionally, the use of a large pinchoff voltage V_P implies that depletion-mode MESFETs are more suitable than enhancement-mode devices, which is in fact the case.

The saturation characteristics of all FETs in general will be softened by the presence of parasitic resistance that appears in series with the device channel. Indeed, this often leads to the failure of experimental devices to saturate. Parasitic resistance in series with the device channel effectively increases the drain-to-source voltage by

$$V'_{DS} = V_{DS} + I_D R_S,\qquad(7.74)$$

which also increases $V_{DS,sat}$ by an equivalent amount. While this series resistance does not change the ultimate value of $I_{D,sat}$ for a given gate bias, it does extend the nonsaturation region, as desired. The use of purposely added series resistance to FETs must be employed with care, because the additional resistance will also decrease the transconductance within the nonsaturated region. In practice, additional series resistance can be achieved by simply reducing the dose of the drain and source implants in a MOSFET, or by increasing the source-to-gate or gate-to-drain electrode spacings of a MESFET.

The saturation characteristics of any FET are also softened by conduction through paths in parallel with the device channel. Parasitic parallel conduction will have the effect of increasing the drain current at a particular bias value,

$$I'_D = I_D + G_P V_{DS}.\qquad(7.75)$$

In practice, such parallel conduction paths arise from conduction through the substrate, or less frequently, along poorly passivated surfaces. In general, these are much more difficult to control than series resistances, and they have the effect of simply adding to the self-relaxation term of the network equations without substantively extending the nonsaturation region. A dynamic range of only 1–2 decades in the node voltage will generally require a strict minimization of parallel conduction paths, unless they are specifically intended to be used as part of the self-relaxation term, which would be an unwise design practice.

7.4 IMPLEMENTATION OF INTERCONNECTIONS

The objective of the design of the interconnection paths in a shunting neural network is to implement the multiplication by the state variable of the receiving cell using a minimum of hardware. A minimum hardware requirement becomes a necessity in any neural network design because the interconnection complexity may be as large as of order $N(N-1)$, where N is the number of cells. This assumes that each of the N cells maintains a unilateral connection with each of the remaining $(N-1)$ cells, although for many applications this degree of interconnection is not required. From an implementation standpoint, minimization of the interconnection hardware still accrues to the overall simplicity of a given design.

As discussed previously, maintaining the multiplicative property of the interconnections may be achieved by preventing the interconnection FETs from reaching a saturating channel current. Assuming that this is accomplished, by a combination of extending the nonsaturation region via device design and limiting the dynamic range of the signal swing via circuit design, the network equations may be phrased in terms of the controlled conductances as

$$C_i \frac{dV_{DSi}^i}{dt} = -G_i V_{DSi}^i - V_{DSi} \sum_{j \neq i} G_{ij}^i(V_{GSi}^i, V_{DSi}^i) - I_i^i$$

$$+(V_{DD} - V_{DSi}) \sum_{k \neq i} G_{ik}^e(V_{GSi}^e, V_{DSi}^e) + I_i^e. \qquad (7.76)$$

The interconnection problem is now to relate the input gate-to-source voltages of each interconnection FET to the state variable of the cell that is controlling it.

The inhibitory interconnections will be examined first. The most attractive method for accomplishing the interconnection is direct coupling, whereby the output node of the jth sending cell is directly connected to the gate of the inhibitory interconnection transistor in the ith receiving cell,

$$V_{GSi}^i = V_{DSj}^i. \qquad (7.77)$$

To first order, the FET controlled conductances are proportional to the gate-to-source voltage minus their threshold voltage. Since the output of the sending cell will be positive definite, this connection will only allow the controlled conductance to be reduced to zero if the interconnection transistor is an enhancement-mode (E-mode) FET. While this introduces additional difficulty for the fabrication of JFETs and MESFETs, an enhancement mode n-channel MOSFET allows direct coupled interconnections to be easily realized in Si process technologies. By virtue of this fact, Si technologies hold the greatest immediate promise for very large scale network integration densities.

For JFETs and MESFETs, which pose process challenges in controlling the threshold voltage of E-mode devices, depletion-mode (D-mode) devices are preferred, also because they usually provide for a greater allowable signal swing. However, to use a D-mode FET as the interconnection transistor requires that a DC level shifting stage be used in between the output node of the sending cell and the gate of the interconnection transistor,

$$V_{GSi}^i = V_{DSj}^i - V_{LS}^i, \qquad (7.78)$$

where V_{LS}^i is the voltage offset for the inhibitory interconnection. Such level shifters can be implemented by the use of a diode stack which is operated with a suitable constant current source. In practice, these typically involve two additional FETs and 2–4 diodes. Although level shifting is required for each unilateral interconnection, greater economy may be achieved by employing only one level shifter per cell, since the output of this level shifter may be fed to the gate input of multiple other cells. This has the advantage of providing some additional output buffering of the sending cell. In order for the level shifter to allow the controlled conductance to fall to zero, the level offset must at least cancel out the magnitude of the threshold voltage of the interconnection transistor,

$$V_{LS}^i > -V_T^i. \tag{7.79}$$

For excitatory interconnections, it is also desirable for the sending cell to be able to drive the controlled conductance to zero. If the interconnection FET is an n-channel device, it is customary to define the drain as the more positive terminal which would then be connected to the V_{DD} positive power supply rail. The source of the device is then connected to the output node. The worst-case situation for this is when the output node voltage is near zero, effectively grounding the source, for which the same requirements as on the inhibitory interconnection apply. Direct coupled interconnections,

$$V_{GSi}^e = V_{DSk}^i - V_{DSi}^i, \tag{7.80}$$

may be used when the interconnection FETs are E-mode types. Again, Si process technology is presently more suitable for this and should allow the highest integration densities to be obtained. Level shifters must be used for D-mode interconnection FETs,

$$V_{GSi}^e = V_{DSk}^i - V_{LS}^e - V_{DSi}^i, \tag{7.81}$$

where V_{LS}^e is the excitatory level offset. This should also be chosen so as to at least cancel the magnitude of the threshold voltage,

$$V_{LS}^e > -V_T^e. \tag{7.82}$$

The presence of a positive state voltage V_{DSi}^i serves to further reduce the drive to the gate of such an excitatory interconnection FET. The implementation of the excitatory interconnection by a FET placed between the positive power supply rail and the output node also provides the circuit with a built-in gain control mechanism. As the state voltage V_{DSi}^i increases, both the drive to the gate of the interconnection FET, $V_{GSi}^e = V_{Gi}^e - V_{DSi}^i$, as well as the voltage across this controlled conductance, $V_{DSi}^e = V_{DD} - V_{DSi}^i$, are reduced. This serves to automatically keep the state voltage of the cell below the positive power supply rail of V_{DD}. The implications of such a gain control function are discussed in more depth in other chapters of this work. For the purposes of interconnection design, it is sufficient to note that the inhibitory and excitatory interconnection transistors have different drain-to-source voltages, their sum is equal to the power supply voltage, and the drain-to-source voltage of the inhibitory transistor is defined as the state voltage of the cell.

Level offset techniques are once again necessary with presently existing JFET and MESFET process technology, and the level shifters for the excitatory interconnections may be constructed in an identical manner to the ones for the inhibitory interconnections. In most cases, a single level shift stage can support drive for all of the necessary inhibitory and excitatory interconnections, so the increase in interconnection hardware is only of order (N).

Si process technology also affords the use of complementary-MOS (CMOS) circuitry, and through this additional options for the excitatory interconnections exist. When a p-channel device is used for the excitatory interconnection FET, the usual convention is to define the source as the more positive terminal and so the drain is connected to the output node of the cell. The n-channel device equations that were developed earlier may then be used for the p-channel case, except that the indices must be reversed, that is, $V_{DS} \rightarrow V_{SD}$, $V_{GS} \rightarrow V_{SG}$, and $I_D \rightarrow -I_D$. The gate-to-source input voltage is thus measured relative to the positive power supply rail. The network equations for a complementary interconnection scheme (n-channel inhibitory FETs and p-channel excitatory FETs) become

$$C_i \frac{dV_{DSi}^i}{dt} = - G_i V_{DSi}^i - V_{DSi} \sum_{j \neq i} G_{ij}^i (V_{GSi}^i, V_{DSi}^i) - I_i^i$$

$$+ (V_{DD} - V_{DSi}) \sum_{k \neq i} G_{ik}^e (V_{SGi}^e, V_{SDi}^e) + I_i^e. \qquad (7.83)$$

If direct coupling is utilized, then

$$V_{SGi}^e = V_{DD} - V_{DSk}^i, \qquad (7.84)$$

which has the effect of inverting the polarity of the interconnection. That is, an increasing state voltage in a neighboring cell will reduce the source-to-gate drive voltage of an excitatory connection thereby reducing the flow of current from the positive supply rail into the output node. Because the controlled conductance is multiplied by ($V_{DD} - V_{DSi}^i$), a gain control function is still implemented by the p-channel FET interconnection. However, the inversion of the polarity of the interconnection means that it may be equally well regarded as an inhibitory interconnection with gain control and with a zero-input steady current flowing. As the state voltage of the sending cell increases, this current flow is reduced. By using a second FET as a current source to offset this zero-input current, an inhibitory interconnection with gain control may be realized. This idea may be extended further by tying the gates of an n-channel inhibitory FET and a p-channel excitatory FET together to form a standard CMOS inverter which is controlled by the state voltage of a neighboring cell. This provides an interconnection with an overall inhibitory polarity, but with gain limiting for state voltages that approach either the positive power supply rail or ground. Level shifters may also be employed to further increase the possibilities available with complementary interconnections, and a large selection of other alternatives also exists, depending upon the specific application.

An important requirement of shunting networks is the need to independently adjust the weights of the interconnection paths. With FET implementations, this problem is easily implemented by simply varying the $\frac{W}{L}$ ratio of the individual interconnection

FETs. In practice, this allows for interconnection strengths varying over a range of 100:1 to 1:100. Larger ratios can be obtained in each direction, but these also demand proportionally more chip real estate and require greater justification for their use. Even so, one can realistically expect approximately four decades over which the interconnection strengths can be varied. The principal drawback to programming the interconnection weights in this manner is that they cannot be adjusted at a later time, and such networks are thus not capable of "learning" processes.

7.5 LIMITATIONS TO PERFORMANCE

In addition to the range and methods for programming the interconnection weights, there are other restrictions on the application of shunting networks that are imposed by the use of FET-based integration technologies. The most principal among these is the degree of interconnection that can be accomodated.

Each unilateral interconnection of the previously described shunting network involves one FET in the receiving cell and one wire connecting the gate of this FET to the output node, level shifter, or buffer of the sending cell. A fully interconnected network thus requires $N(N - 1)$ unilateral inhibitory interconnections and $N(N - 1)$ unilateral excitatory interconnections for N cells. Because the output buffer or level shifter of a given cell may serve to drive all of the interconnections, only N of these are required. Similarly, the space occupied by the interconnection wires arises mostly from the path between two cells and a single run may be used for both the inhibitory and excitatory interconnections by simply connecting to the gates of both of these FETs at the receiving cell. Thus the maximum number of interconnection transistors is $2N(N - 1)$ while the maximum number of interconnection wires is $N(N - 1)$. The required fanout from the output of each buffer or level shifter is $(N - 1)$. For small networks of order $N < 100$, none of the above quantities are prohibitive to the fabrication of such circuits in monolithic IC form. However, as the number of cells in the network increases, the number of wires and interconnection transistors which are of $O(N^2)$ complexity will severely limit the maximum network size that is achievable with this architecture. As the number of cells increases above approximately 100, the chip area dedicated to the interconnection wires will begin to exceed the area used to implement the network cells themselves. Such a situation is always a limitation in a planar integrated circuit. For fully interconnected networks of this type, an upper limit of around 10^3–10^4 cells is anticipated.

Many important applications for neural networks exist in the processing of sensory input data, such as visual or auditory signals (Mead 1989). These networks become much more practical for planar integrated circuit implementations because the degree of interconnection only involves the nearby neighbors of a given cell. If each given cell is required to connect to only M of its nearest neighbor cells, then the number of interconnection transistors per cell is reduced to $2M$, the fanout of each cell is reduced to M, and the number of interconnection wires is reduced to NM. Such networks are easily implemented with M of around 10 or less. Additionally, since the area required by each cell and its interconnection wires is now a constant, the area of the overall network scales linearly with the number of cells. The maximum number of cells achievable with such a nearby neighbor interconnection matrix is limited only by the overall size of the die and with the presently existing technologies is approximately 10^4 cells for JFETs or MESFETs and approximately 10^5 for MOSFETs.

Because no gate current flows under normal static operating conditions, FET-based technologies are not subject to restrictive fanout limitations in so far as their DC characteristics are concerned. For switching purposes, however, increasing fanout will increase the capacitance of the output node and will reduce the bandwidth of the circuit. This capacitance will directly add to the existing node capacitance C_i, and will influence the dynamic characteristics of the network. Because it is not practical to construct large node capacitances directly on the IC, the capacitance across which the state variable of voltage appears will be determined largely by parasitic node capacitances which are outside of the network designer's ability to control. Unless the designer is able to externally load each output node with a large capacitance, he will simply have to live with whatever capacitances the specific fanout and wire routing provide.

Power dissipation is a more serious issue with shunting networks. The current flow through an individual cell is the sum of the input currents plus any excitatory currents. In a static situation, this is obviously equal to the sum of the inhibitory currents plus the self-relaxation current. The voltage across each cell is V_{DD} which is a constant, so the power dissipation of each cell is proportional to the degree of its excitation. The design of the overall chip must allow for the worst case of each cell operating at maximum current flow. For a typical positive power supply voltage of $V_{DD} = 5$ Volts and a maximum cell current of 1 milliampere, a total power dissipation of 5 Watts per 1000 cells becomes a significant design issue. This figure will be substantively increased by the presence of buffer or level shifter stages.

The input to each cell in a shunting network is a current and the output state variable is a voltage. Both of these impose constraints on the allowable dynamic range of the network. The maximum value of the state variable V_i is equal to the positive power supply rail of V_{DD}. The minimum value is set by the noise level of the node which is usually a combination of thermal and shot noise sources. Additionally, $GaAs$ circuitry exhibits a 1/f noise corner frequency that is approximately 10^3 higher than that of equivalent Si circuits. The contribution from these depends strongly upon the bandwidth of the node B and the real part of the nodal impedance R. For the thermal contribution, the *rms* noise voltage contribution is

$$V_{rms}^2 = 4kTBR. \tag{7.85}$$

In practice, the minimum signal levels are approximately 20-30 millivolts, which entails an output dynamic range of only 2.5 decades.

For the input currents, the maximum value is set by the power dissipation requirement discussed previously and may be as high as 10 mA for some designs. The minimum value is set by the leakage current level for the FETs. For $GaAs$ MESFETs, the leakage current level is caused by conduction through the semi-insulating substrate and leakage through the reverse-biased gate, and both are typically on the order of 1 microampere. For Si MOSFETs the leakage current is much lower, typically around 1 picoampere (Taylor 1978). In both cases, the FETs are operating in the subthreshold regime. These current levels produce an input dynamic range of approximately 4 decades for $GaAs$ JFETs or MESFETs and approximately 10 decades for Si MOSFETs. Thus, Si offers a considerable advantage for networks that must operate over such a wide dynamic range of inputs, but the output dynamic range is still roughly equal for both materials choices. Since the dynamic range of the output is much less than the dynamic range of the input, the network must achieve some form of signal compression in order to make the fullest use of both ports. As has been discussed in other parts of this book, shunting networks inherently produce a square-root signal compression factor which makes these networks attractive for implementation in either materials system.

Acknowledgements.   This work was supported by the National Science Foundation under grant no. MIP-8822121 and by the Washington Technology Center under project no. 09-1091.

8

Specific Implementation

The simplest shunting recurrent neural network that can be implemented within the design framework of Fig. 6.2 and demonstrates the desired properties of shunting networks is shown in Fig. 8.1. This network implements the network equation

$$\frac{dx_i}{dt} = I_i - a_i x_i + K_i x_i^2 - x_i \sum_{j \neq i} f_j(x_j). \tag{8.1}$$

This circuit was implemented using depletion-mode Junction Field Effect Transistors (JFET's)[1] due to their similar characteristics to gallium arsenide MESFETs. This implementation thus can serve as a precursor to a GaAs-based monolithically integrated circuit implementation of an optically-sensitive network, and as a vehicle for testing the design principles and circuits presented in the previous chapters.

The simplicity of the design should be stressed, since in this realization:

- Each "neuron" consists of only one resistor (transistor T_1), in parallel with a cell capacitance, C.

- Each multiplicative coupling is done by *one* transistor (T_{ij}).

- Transistors T_2 and T_3 and diode D_1 form a level shifter which is necessary for correct voltage level interfacing in d-mode technology but not required when using e-mode devices.

[1]Motorola, MPF102, depletion-mode JFET.

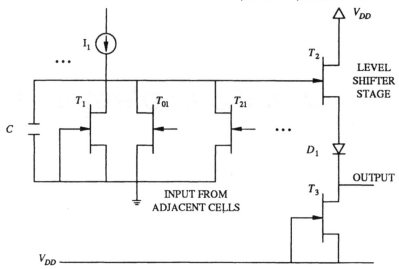

Figure 8.1 Implemented recurrent shunting network using d-mode FET's.

- Input can be directly supplied with one Metal-Semiconductor-Metal (MSM) photodetector. In the discrete circuit implementation one transistor, with a variable resistor applied to its gate and source terminals, was used as current input.

A two-dimensional matrix with 4 nearest neighbor interconnections will then consist of only 7 transistors, 1–3 diodes depending on the amount of level shifting required and the turn-on voltages of the diodes, and if desired, 1 capacitor per cell. The compactness of the design is notable.

A circuit consisting of 7 cells with nearest neighbor connections was constructed. The following sections describe important design considerations and the experimental results obtained from this implementation. All the results favorably compare to results of SPICE simulation of the network.

8.1 ACCURACY

Since the "workhorse" of the present implementation is the operation of a FET in the subsaturated linear region, it is important to verify the validity and accuracy of the simple mathematical model of the operation of a FET in the linear region, namely Eq. (6.2), which is repeated here

$$I_{DS} = K \left[(V_{GS} - V_{th})V_{DS} - \frac{1}{2}V_{DS}^2 \right], \quad \text{for} \quad V_{DS} < V_{GS} - V_{th}. \quad (8.2)$$

Also, often in modeling the operation of a FET in the linear region the quadratic term $\frac{1}{2}V_{DS}^2$ in Eq. (8.2) is ignored and the transistor is modeled as a linear voltage-controlled conductance. Since such an assumption directly alters the network equations, its applicability should also be experimentally determined.

The device-dependent parameters K and V_{th} in Eq. (8.2) were extrapolated from experimental data and used in linear and nonlinear models of a transistor. Comparison of the models with experimental data showed that ignoring the quadratic term produces an average error of 25% while including this term results in a much lower average error of 7%. For this reason the model of (8.2) which includes quadratic terms will be used throughout this work.

The previous experiment also implicitly shows the *accuracy* of the performed multiplication and is better demonstrated by the following experiment.

The simplest tractable network equation can be found by applying the output of a cell to itself. This is mathematically equivalent to the application of a *uniform input* to an arbitrarily large array of identical neurons and hence does not constitute a trivial example. Such a system is described by

$$\frac{dx_i}{dt} = I - ax_i + K_1 x_i^2 - K_2 x_i x_j,$$ (8.3)

where $x_i = x_j$ for all j.

This equation has a straightforward analytical solution that can be compared with experimental data to show the accuracy of the model and thus provide a precision measure for the multiplication performed as well as the network equations implemented. Figure 8.2 shows the result of such a comparison, where it is seen that the equations implemented are accurate to within an average error margin of 1.89%.

Another figure of merit that helps put the accuracy of the implementation in relative perspective is found by noting that the 1.89% average error can be treated as quantization error of a corresponding digital system. A six-bit digital code produces roughly the same average error *if* the output falls within the range (1–64) that can be coded by six bits. However, the range of the output, and input, far exceeds (1–64) and a digital system which is capable of addressing even the limited range of Fig. 8.2 requires at least 9 bits to represent this range. Based on these considerations, a digital 8-bit machine would be needed to perform this computation with equal accuracy.

8.2 LEVEL-SHIFTING AND THE DYNAMIC RANGE

Transistors T_1 and T_2 and diode D_1 in Fig. 8.1 perform the operation

$$V_{out} = V_{in} - V_{LS},$$ (8.4)

which shows that the output is a version of the input shifted by the amount V_{LS}, hence the name level-shifter. Experimental results of a level-shifting stage with three diodes[2] is shown in Fig. 8.3, where it is seen that the output is described by

$$V_{out} = V_{in} - 1.90$$ (8.5)

in agreement with Eq. (8.4).

[2]Motorola IN4007 was used.

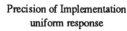

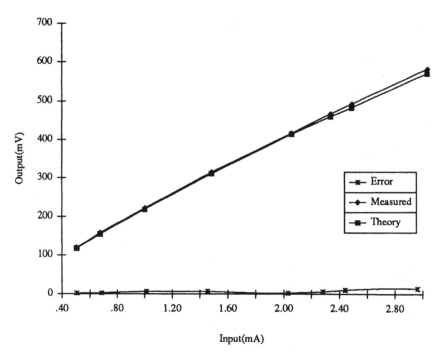

Figure 8.2 Response of a cell with self-inhibition; comparison of experimental and theoretical results, showing the accuracy of the implementation. Average error is about 1.9%.

The amount of level-shifting is only a function of the turn-on voltage of the diode D_1 which can be increased, up to the negative supply rail voltage, by adding more diodes in series, or custom designing diodes with required turn-on voltages during integrated circuit design. This amount is a crucial design parameter since it establishes the dynamic range of operation, as explained below.

The output of the level-shifter is applied to the gate of the d-mode transistors T_{ij} (Fig. 8.1) of other cells in the network and controls the conductance of these transistors. It should thus be smaller than voltages that may cause excessive current flow by forward biasing these gates. On the other hand V_{out} is lower-bounded by the threshold voltage of these transistors, since if it is reduced below the threshold voltage, the conducting transistor channel will be completely pinched-off and the transistors will act as open circuits. Since the voltage of each cell is the input current multiplied by the equivalent resistance seen by it, elimination of a parallel conduction path increases the resistance and thus the cell voltage unproportionally. This error is more significant for networks with fewer interconnections.

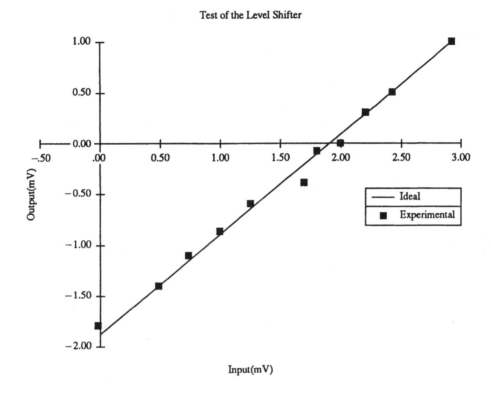

Figure 8.3 Experimental verification of level-shifting stage.

From the above discussion, the level-shifting amount should be chosen within the range

$$V_{th} \leq V_{LS} \leq V_{up}, \tag{8.6}$$

where V_{up} is a positive quantity, typically a few tenths of a volt and V_{th} is the negative valued threshold voltage. Furthermore, in this range an operating point which maximizes the possible conductance swing is optimal.

In the circuit of Fig. 8.1, the threshold voltage of the transistors was determined to be 1.5 volts, thus one diode with a turn-on voltage of about 0.6 volts was chosen to provide the optimal level-shifting amount.

Finally, in addition to providing correct polarity, the level-shifter stage performs an important interfacing role. Since the input to the level shifter is applied to the gate of transistor T_1, a reverse-biased diode for MESFETs, this stage does not draw any significant amount of current and practically isolates the output of a cell from other cells it is connected to and thus allows for a large fan-out to be achieved. The level-shifter does conduct current and limits the size of implementation by its power dissipation requirements.

The allowed range of the *input*, on the other hand, is dependent on the resistance seen by the input and is determined by the pinch-off voltage of the transistors. This is best demonstrated by an example.

Consider a cell with the typical[3] membrane resistance of 400Ω, connected to four neighbors. The highest resistance seen by such a cell occurs when all the interconnect transistors have their highest resistance value, which occurs for $V_{GS} = V_{th}$. For the transistors used, this value equals 2K ohms, and hence the equivalent resistance is

$$R_{eq} = 400\|2000\|2000\|2000\|2000 = 222\Omega.$$

The cell voltage, which is the drain-to-source voltage of parallel transistors, is found from

$$x_i = I_i \times R_{eq} \tag{8.7}$$

and should remain in the below saturation region. For a long channel transistor this range is a few volts and hence the input I_i in (8.7) can rate up to 10–15 milliamperes.

It should be noted that even if the input current exceeds this range, the network remains stable, only the shunting interaction is replaced by an additive interaction for the specific cell, which due to its high activity strongly inhibits its neighbors and forces them deeper into below saturation operation. In practice, power dissipation requirements determine the upper-bound for current flow in the network.

As seen from this qualitative treatment, the higher the degree of interconnection, the larger the dynamic range within which the input can be processed. The allowable *lower bound* of the input is determined by the sensitivity of the devices and ultimately by the noise immunity of the technology. With present technology it seems quite feasible for several orders of magnitude of the input current, from high nanoamps to low milliamperes, to be processed. The wide dynamic range achieved from a limited dynamic range of each unit is a good example of the power of collective behavior of neural networks.

8.3 RESPONSE TO UNIFORM INPUT

As mentioned previously, the response of a network of identical cells to a uniform input can be deduced from the response of one cell to a single input since a uniform input is only characterized by an intensity level without any contrast, or pattern. A uniform input response is thus significant in demonstrating the intensity dependent properties of networks.

Figure 8.4 shows the response of a network to a uniform input and is seen to be acceptably uniform across the array, as expected. The lack of uniformity, noise, that is observed in this experiment can be explained by the device variations across the array and is seen by observing that the simple level-shifter stage which only consists of 2 transistors and 1 diode varies by up to 68 millivolts across the array, while the uniform input response varies by at most 17 millivolts across the network. Furthermore, the

---

[3]Typical for the devices used, not the biological units.

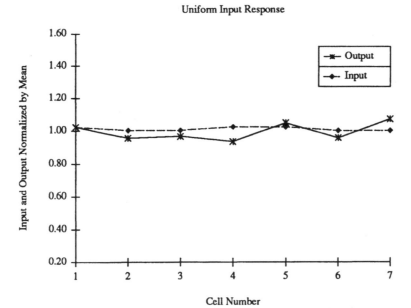

Figure 8.4 Response of the network to uniform input; variation across the array is mainly due to component variation.

abrupt boundary influences the network response and produces errors which should be accounted for. In all of the experiments a constant load was added to cells number one and seven to compensate the abrupt boundary. More sophisticated techniques were employed for cases where boundary effect cancellation was crucial and will be shown as they are encountered. Results of SPICE simulation show a completely uniform response, as expected.

Having established that the uniform input response can be found from the response of one cell to an input, from Eq. (8.3), it is seen that the response of the network to a uniform input is proportional to the *square root* of the input but is also a function of the parameters K_1 and K_2 which are dependent on the membrane resistance and transconductances of the devices respectively. Figure 8.2, which shows the response to a uniform input, is seen to be close to a linear function, but can be tuned, by a different choice of K_1, to show the square root dynamic range compression. The mechanism and uses of this tuning process will be investigated in a separate section. Figure 8.5 shows the response when the membrane resistance value is chosen to better reflect the desired slower-than-linear characteristics of the network.

Figure 8.5 is drawn from actual implementation, this behavior is more pronounced when the neural network *equations* are analyzed independent of the implementation constraints. Figure 8.6 shows the steady state response from a theoretical analysis, presented in Chapter 9, of a network consisting of 3 fully interconnected cells, where the coefficients are chosen to better demonstrate the dynamic range compression property.

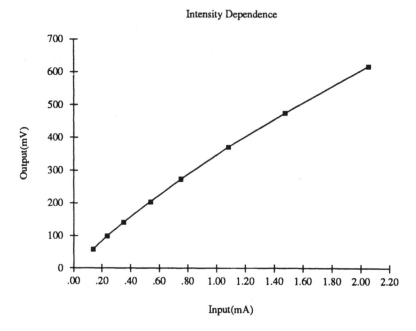

Figure 8.5   Response of the network to uniform input. The output is proportional to the square root of the input.

8.4 RANGE COMPRESSION AND DATA COMPRESSION

Figures 8.5 and 8.6 show the dynamic range compression *beyond* the sensor layer, which for most biological systems has already performed a logarithmic transformation of the input. The significance of the range compression property is seen by noting that the human retina, for example, is sensitive to an input range of 1–10^{13} luminance units (Levine 1985) while the firing range of the ganglion cells, whose axons form the optic nerve, is 1–200 spikes per second, and is further reduced by the presence of intrinsic noise. Clearly, the compression of the dynamic range of the signal beyond the photoreceptor layer is essential in achieving such a wide dynamic range of sensitivity without saturating the processing power of each cell.

This property has immediate technological applicability; presently available photodetector imaging technologies such as charge coupled devices (CCD) arrays cannot compare with the sensitivity range of the biological systems without the attendant degradation caused by blooming, ghosting, or minimum charge transfer sensitivity.

Range compression, however, is of little value if it is not concomitant with other processing capabilities, and especially if important features of the data are not preserved. The following observations lead to the conjecture that the network performs *data compression* in the information theoretic sense of the removal, or at least reduction, of redundancy and not simply range compression.

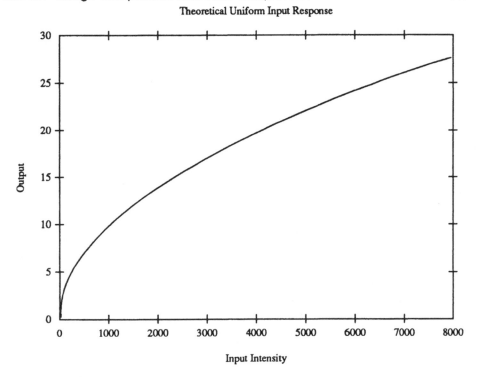

Figure 8.6 Theoretical response of the network to uniform input showing the square root compression of the input range.

- Spatial edges, or areas of contrast, contain the most information while uniform areas are predictable (redundant) and contain little information. The future sections provide experimental data that the network contrast-enhances the input while attenuating uniform areas. In general, a center-surround anatomy automatically decreases redundancy via transduction of the differences but it may also introduce unacceptable distortion.

- The auto-correlation function of the output tends to be narrower than that of the input, that is, the output is less correlated than the input, and hence contains more independent data points, more information, than the input. This is a crucial test but needs to be verified by a larger pool of neurons since all statistical measures are less meaningful for such short-extent sequences.

- The experimental data is very similar to measurements of insect visual cells, which gives rise to the hypothesis that the peripheral visual system performs a predictive coding of the input data (Srinivasan et al. 1982) resulting in the removal of redundancy. Shunting networks seem to be capable of mediating the same process, with many other properties that can not be explained by a predictive coding hypothesis. This again needs to be verified by a larger number of neurons than the present discrete component implementation and is a goal of the integrated circuit implementation.

It should be noted that data compression, if rigorously established, is only *one feature* of the shunting nets rather than their main purpose. Maximum entropy coding techniques, cost functions, and error criteria, as well as other conventional processing techniques can act as comparative measures of performance for specific tasks while the neural network theory should be distinct and more general than any single conventional hypothesis.

8.5 POINT SOURCE RESPONSE

Response of a neural network to a point source input is important in characterizing the behavior of the network, similar to the manner in which the impulse response characterizes the behavior of a linear system. There are, however, distinct differences. The temporal impulse response, or the spatial point-source response, completely specify the temporal properties of a linear system and its Fourier transform provides the temporal frequency transfer function, or the spatial frequency modulation transfer function, which completely specifies the frequency domain properties; the same principles do not apply to nonlinear systems such as the shunting networks.

The point-source response can be assumed to specify the "receptive field" of a *stimulus*. This assumption requires some clarification especially since the results will be compared to the receptive field of biological cells.

The phrase "receptive field" was coined by Hartline (1940), who solved the problem of dissecting a nerve bundle in the retina of the frog and noted that no description of optic responses in single fibers would be complete without the description of the region of retina that must be illuminated in order to obtain a response in any given fiber. Hartline called this region a *receptive field*. Receptive field is thus a map of the spatial sensitivity of a *cell* to stimulus. Classic works by Kuffler (1953), Hubel and Wiesel (1959), and Barlow (1962) carefully described the shape of the receptive field of the cat ganglion cells, the cells in the lateral geniculate nucleus of the cat, and rabbit ganglion cells, respectively, which helped explain the response of these cells to different stimuli.

On the other hand, a stimulus, a narrow spot of light for example, invokes a pattern of activity across the cells; it excites cells on whose on-center it has fallen and inhibits those whose off-surround receptive field contain the stimulus. This pattern of response is taken here to specify the receptive field, even though a "projective field" would have been a more appropriate terminology, and is strictly valid when the cells have identical receptive fields, which is the case for the present implementation.

As a point of reference, the response of a circuit *without* any interconnection, that is, without network interactions, is shown in Fig. 8.7 where, as in all the experimental results reported here, data is valid only at cell locations as indexed by the cell number and linear interpolation between data points is for visual consistency only. The sensory array reproduces the input without processing it; as the case would be for most conventional sensing apparatus such as CCD camera arrays. This figure also shows that noise is

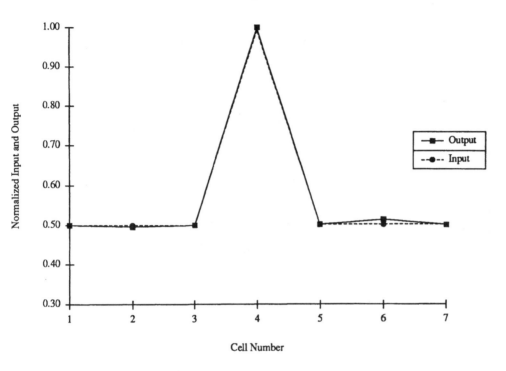

Figure 8.7   Input-output relationship without a neural network; network traces the input without processing it. Both input and output curves are valid only at cell numbers, the linear interpolation between data points is only for visual consistancy.

insignificant and the global interaction of the network has countered small parameter variations such as variation of the level shifting stage among cells.

The response of the network to a point source input is shown in Fig. 8.8. The on-center off-surround shape of the response is similar to those observed in many excitable cells including the works mentioned above (Kuffler 1953; Hubel and Weisel 1959; Barlow et al. 1964). Inset of the figure (from Pinter et al. 1990) shows the receptive field of the Lamina Monopolar Cells (LMC) of the fly *Lucilia sericata* and is derived by measuring the response of the fly to gratings of different spatial frequencies, which provides a Modulation Transfer Function (MTF), curve fitting the data to find an analytical expression for the MTF, and inverse fourier transforming this expression.

This type of receptive field has been approximated (Marr and Hildreth 1980, Marr 1982) by the difference of two gaussian functions, a narrow excitatory center and a wide inhibitory surround, and is shown to be very similar to a gradient (laplacian) operator. Important properties of such receptive fields have been well documented and have given rise to a prominent school of thought in computational vision known as the "zero crossing

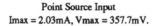

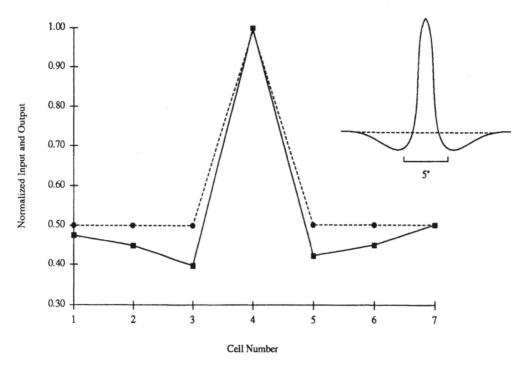

Figure 8.8 Response to a point-source input. Inset shows the receptive field of LMC cell of *Lucilia sericata*. Horizontal axis of inset in visual angle, vertical axis relative voltage units of hyperpolarization. Inset from Pinter et al. (1990).

school" since the second derivative of an (spatial) edge which is smoothed by a gaussians kernel is zero at the location of the edge and thus the zero crossings of this function specify the edge locations.

8.5.1 Intensity Dependence

The zero crossing computation is a feed forward linear operation, and although very effective it does not posses the computational power or the versatility of the nonlinear shunting nets, specifically any of the properties shown in Section 3.2. One such property is adaptation to mean intensity level and is demonstrated in Fig. 8.9.

It is seen from this figure that as the background intensity increases the receptive field becomes more pronounced, the inhibitory flanks become relatively stronger showing higher levels of network interaction. The profile of the receptive field thus changes according to the intensity level. The inset shows a strikingly similar behavior in *Lucilia* which proves that shunting nets are capable of mediating the receptive field adaptation to the mean-intensity level observed in many biological units such as the LMC cells of

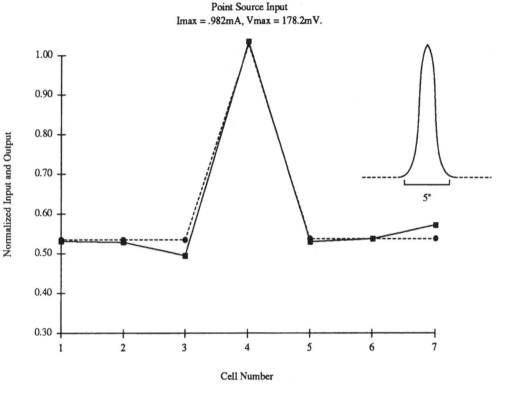

Figure 8.9 Adaptation of the receptive field to mean intensity. Inset shows the receptive field of LMC cell of *Lucilia sericata* at lower luminance levels. Inset from Pinter et al. 1990.

Lucilia. This is consistant with theoretical analysis of multiplicative inhibitory networks (Pinter 1983b, 1984).

If a modulation transfer function is defined as the fourier transform of the receptive field, comparison of Figs. 8.8 and 8.9 shows that as the mean intensity increases the MTF extends to higher spatial frequencies, that is, the network *sees* better, with more detail, in higher light levels. This result has also been theoretically shown by Pinter (1983b).

Intensity dependence is a very important property; it shows a "retina" that uses available ambient light to better process the input. This provides *one* explanation for why a light switch is turned on to see better in low luminance levels even though the reflectances, the percentage of the light reflected from objects, are independent of the amount of light. The same applies to other sensory systems, a speaker is asked to speak louder to become more intelligible, even in the absence of noise, while the spectrogram of the speech does not change when the signal is amplified. This property is a result of the reflectance-luminance interaction that was shown in Chapter 3 and is specific to shunting nets.

Figures 8.8 and 8.9 have been normalized to better show their relative shapes and are clearly distinguishable even after normalization. Considering that the only difference between these two patterns is their background intensity, the network has in fact *coded the intensity information* as well. The difference between the two figures provides a clue to the amount of the intensity. This is lost in most front-end processors which require carefully regulated lighting conditions or sensors whose output has to be normalized prior to any processing so as not to exceed the dynamic range of processing circuitry. The luxury to encode both the pattern and its relative intensity is due to the unique wide dynamic range capability of shunting nets.

8.5.2 Boundary Effects

The asymmetry of the receptive field, especially the difference between the response of cells 1 and 7 is due to component variation as well as the abrupt boundary. The effect of the boundary has been somewhat compensated by application of a dummy load to cells 1 and 7, but can be further reduced by connecting these cells together. Such a connection imposes a cyclic boundary condition on the network which is equivalent to the application of a periodic replication of the input to an unlimited array of cells. The result is shown in Fig. 8.10 and is clearly much more symmetric than Fig. 8.8.

8.6 TUNABILITY OF SENSITIVITY

The sensitivity of a network to a set of parameters can be defined as the amount of change that is caused by the variation of these parameters. For the present network the sensitivity of each cell can be changed, or tuned, by varying the membrane resistance. This is seen by noting from Fig. 3.1 that each cell's output equals the product of the input and the resistance seen by the input, that is,

$$x_i = I_i(R_0 \| R_{eq}), \tag{8.8}$$

where R_0 is the membrane resistance and R_{eq} is the equivalent resistance of the conductive paths established by multiplicative interconnects. As a first-order approximation, that is, the first term in a Taylor series expansion, the sensitivity of the network to variation of the membrane resistance is found from

$$\frac{\partial x_i}{\partial R_0} = I_i \frac{R_{eq}^2}{(R_0 + R_{eq})^2}, \tag{8.9}$$

which is a positive definite quantity. Hence as the membrane resistance is increased the output is increased, that is, the cell shows more activity even though the input and interconnections have not varied. By changing the membrane resistance of all the cells in the network, the network can be tuned to be more, or less, sensitive to the environment. This is the tunability property and is experimentally shown in Fig. 8.11, where the input pattern has remained constant but the network is tuned to be more sensitive. It is seen that the response shows a clear change of activity pattern; the inhibitory flanks are more

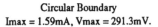

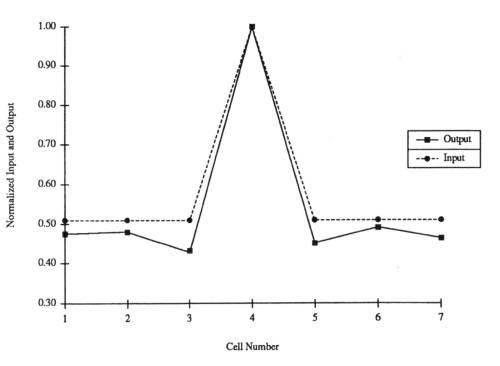

Figure 8.10 Circular boundary condition reduces the boundary effects and adds to the symmetry of the response.

pronounced and the contrast enhancement and gain have both increased. This change of activity is desired when the total input incident on the network is small and there is no danger of saturating the dynamic range of the output neurons. For an optical input case, the lower curve is desired for operation in dim light conditions.

In the present hardware implementation, the tuning is simply accomplished by changing the bias on the transistor which models the membrane resistance. Figure 8.11 shows how the image sensor can be tuned to operate optimally in bright or dim lighting conditions. The prospects of using this property in a self-regulating mechanism which lets the device adapt its behavior according to the environment it is operating is very promising and is presently under study.

It seems possible that such a natural mechanism be employed in biological systems, that is, membrane resistances may be self-modulated to adapt to sensitivity requirements. Verification of this prediction and identification of the mechanisms that may mediate such a process is highly desirable.

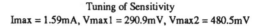

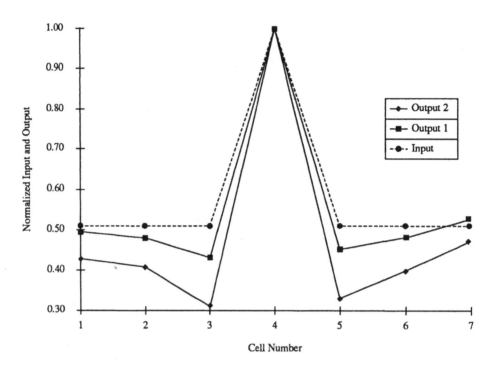

Figure 8.11 Tunability of the sensitivity; the same input pattern produces different behavior when the network is externally tuned (lower curve) to be more sensitive.

8.7 SPATIAL EDGE RESPONSE

Edges, boundaries, and areas of high contrast specify important features of a scene. They also provide the areas of highest information. The response of the network to a spatial edge pattern is shown in Fig. 8.12; edges are enhanced and uniform areas are suppressed. Inset of the figure is the response of the network if no interconnections are made, and is intended to be contrasted with the network behavior. In addition, the inset shows that variation of individual cell properties due to device nonuniformities does not considerably affect the response.

The cells that are located next to the edge location also show an interesting response; they are less active than the next-nearest cells even though they receive equal inputs. This phenomenon has been observed in psychophysical experiments, has been studied by Ernst Mach (1838–1916), and is named *Mach bands* in his honor (Ratliff 1965). The seminal works of Hartline on the lateral compound eye of *Limulus*, which led to a Nobel prize, shows the same pattern of response and is shown in Fig. 8.13 for comparison.

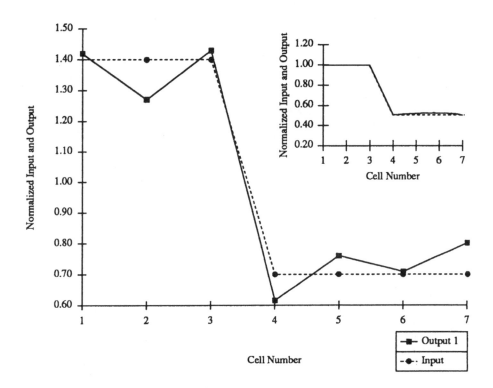

Figure 8.12 caption and chart labels:

Spatial Edge Input
Mean Input = 1.287, Mean Output = 217.23

Figure 8.12 Response of the network to a spatial edge pattern; edges are expanded and Mach bands are observed. Inset shows the response without network interaction.

The importance of edge enhancement for image processing applications is well known, the simplicity of the design, the parallel construction of the network, and its operation in the below saturation region can make the GaAs based implementation of this circuit among the fastest edge detectors. It should be reminded that edge enhancement is only one feature of the diverse capabilities of this network.

A striking demonstration of the importance of the response of the type shown in Figs. 8.12 and 8.13 is provided by Stockham (1972) whereby a filter cancelled the preprocessing accomplished by the peripheral visual system, such that the output of an edge pattern would "appear" as the input itself (as in the inset of Fig. 8.12). It was shown that such an image would be completely smeared and unrecognizable.

The test edge pattern used in this section is a one-pixel edge, that is, the areas of high- and low-contrast are separated by one cell. Often edges are ramps which cover several intermediate cells. There is strong correlation between the degree of interconnec-

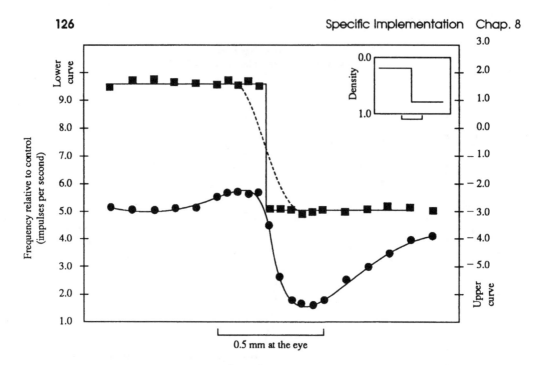

Figure 8.13 Response of the eccentric cell of the lateral eye of *Limulus* to a spatial edge pattern. The upper curve shows the response of one ommatidium only while the lower curve shows the effect of network interaction. Inset shows the spatial input pattern. From Ratliff and Hartline (1959). Reprinted from the Journal of General Physiology, 1959, Vol. 42, No. 6, pp 1241–1254 by copyright permission of Rockefeller University press.

tions and the slope of the ramp, or the edge depth, which is preferentially enhanced by the network; relevant results to the study of directional selectivity will be mentioned in that section. For edges of fixed depth, it is shown in Darling et al. (1989) that interconnectivity beyond the nearest neighbor range produces a stable, geometrically decreasing oscillatory skirt around each spatial edge input. Decay rate of this spatial oscillation is also shown to be dependent upon the contrast and intensity of the input edge and the membrane resistance.

     Not all areas of contrast are, however, edges. Figure 8.14 shows a more complicated test pattern than that of Fig. 8.12, where it is seen that the network is still capable of contrast-enhancing the pattern. It was experimentally determined that when the pattern variation approached the pixel level, the network treated the input as noise, as expected, and was incapable of establishing a correlation between different data points.

     Furthermore, it was determined that the smallest contrast that produced a perceptible edge expansion was 1.1:1, any edge pattern where the ratio of the light side to dark side was less than this ratio produced a noisy output pattern. This ratio may be improved by insertion of a sigmoidal nonlinearity in the feedback loop and optimizing the slope of this nonlinearity for the background input level.

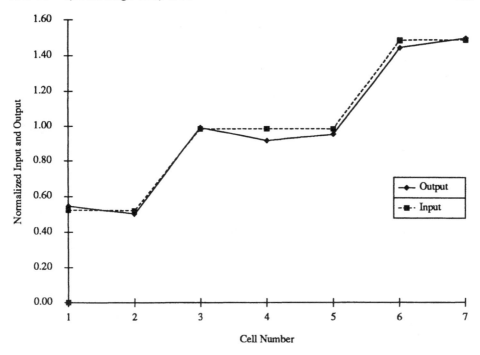

Figure 8.14   Contrast enhancement property.

8.7.1 Intensity Dependence

Adaptation to mean intensity level which was shown for the point source input can also be observed for the spatial edge pattern, as demonstrated in Fig. 8.15. This figure has the same contrast as Fig. 8.12 but a smaller mean luminance level. The network is still capable of expanding the edges but the amount of expansion is reduced at dim light levels. Inset of the figure shows the response of *Limulus* to a *limited extent* spatial edge pattern and is chosen to stress the similarity at the boundaries.

A family of curves shown in Fig. 8.16 demonstrates the adaptation to mean intensity level of the input. As the intensity increases the edges are expanded further, the network thus "sees" better when more light is available. Enhancement of the edges in the space domain is translated into higher spatial frequencies in the frequency domain. This figure thus shows that the modulation transfer function is extended to higher frequencies as the input increases, that is, the network gains more bandwidth, or sees more detail, as more ambient light, or more total activity, is provided. The tuning of the MTF, and its application to vision has been studied in detail by Pinter (1984, 1985), and is hereby experimentally verified.

Behavior of each of the 7 cells of the network in response to a change in intensity is plotted in Figs. 8.17 and 8.18. The response of each cell is seen to be both a function of the input strength *and* the reflectance pattern. Cells 4 and 5, for example, receive

Spatial Edge Input
Mean Input = .386mA, Mean Output = 69.85mV

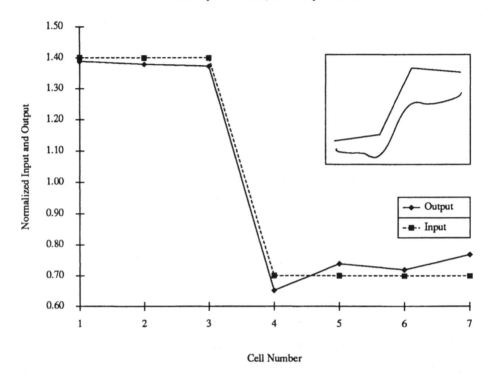

Cell Number

Figure 8.15 Response to a spatial edge pattern with less mean intensity than Fig. 8.12. Inset shows response of the eccentric cell of the lateral eye of *Limulus* to a limited extent edge input. Inset from Ratliff and Hartline (1959). Reprinted from the Journal of General Physiology, 1959, Vol. 42, No. 6, pp. 1241–1254 by copyright permission of Rockefeller University press.

the same input but have drastically different responses to change of intensity. The cells which most actively participate in processing of the reflectances, most conservatively respond to intensity change and thus preserve their dynamic range. This is consistent with the previously mentioned brightness contrast: increase of activity of a population occurs only at the cost of the reduction of the activity of another population.

Figure 8.16 and Figs. 8.17–8.18 complement each other; in Fig. 8.16 the intensity dependence of reflectance processing is stressed and in Figs. 8.17–8.18 reflectance dependence of response to intensity is shown. All three plots are drawn from the same data, which once again proves the interdependence of reflectance and intensity and demonstrates the capability of shunting nets to capture some of the most salient features of peripheral visual processing.

Figures 8.17 and 8.18 are, as expected, similar in shape to the uniform input response which was used to show intensity dependence properties of the network. Had

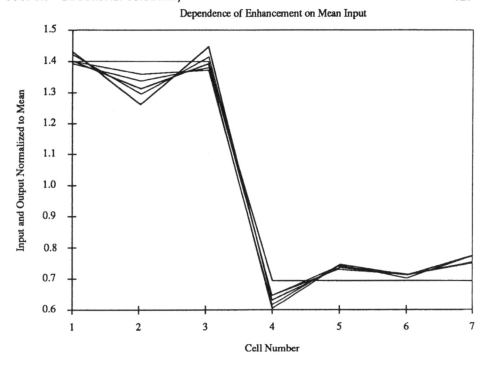

Dependence of Enhancement on Mean Input

Figure 8.16 Intensity dependence of the response is increased from top to bottom. All curves are normalized by mean intensity.

the circuit been tuned to higher sensitivity, the curves would have shown better slower-than-linear response.

8.8 DIRECTIONAL SELECTIVITY

The network can be wired, by choosing asymmetric interconnections, to be preferentially responsive to an edge of a certain direction over other orientations. Figures 8.19 and 8.20 show response of a network which has only right going inhibitory connections, as the extreme case of asymmetry, to edge inputs of equal contrast and intensity but different directions. These figures are normalized by the same values so as to preserve the relative shape of the response. The boundary effects are also crucial since if cell 1 does not receive input from any other cell it will have an unproportionally high value. The same applies to cell 2 which receives input only from cell 1. If these boundary effects are not compensated they will dominate the response characteristics. The solution to adverse boundary effects was provided by applying the output of the cell 1 to itself and cell 2 such that input would "appear" to be extending from the left.

As expected, Figs. 8.19 and 8.20 show drastically different responses to edges which only differ in orientation. The difference of response is both in terms of the shape

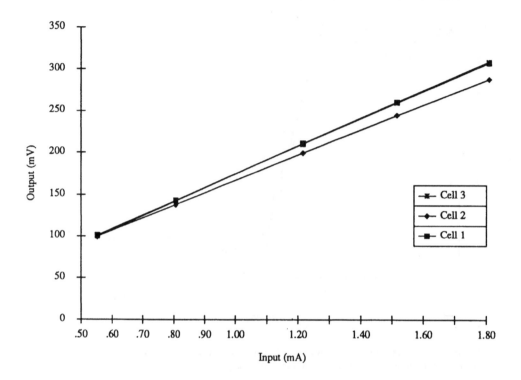

Figure 8.17   Intensity dependence of the response of each cell.

and intensity and hence orientation can be inferred from either. An edge which steps up from left to right, a right handed edge (Fig. 8.19), is contrast enhanced in the usual manner, while a left handed edge is smeared and attenuated. Observation of the operation of cell 4 alone can, in principle, determine the orientation of the edge.

In response to a left handed edge, cells which receive higher intensity can strongly inhibit other cells and thus attenuate the total activity of the network. Total activity of Fig. 8.19 differs from that of Fig. 8.20 by 119 mV for two nearest neighbor connections and by 105 mV for nearest neighbor connections. Hence by monitoring only the total activity of the network it is possible to distinguish between two edges of same contrast and intensity but different orientation; the network is more "agitated" when an input with its preferential direction is presented.

Motion detection of an object within an image can readily be calculated by subtracting two edge-enhanced image frames separated by some short time interval (Bouzerdom and Pinter, this volume, Ch. 5). In a directionally sensitive network, an object moving to the left will produce a greater difference and will be more easily identified. Preferential directional sensitivity thus will result in preferential motion sensitivity.

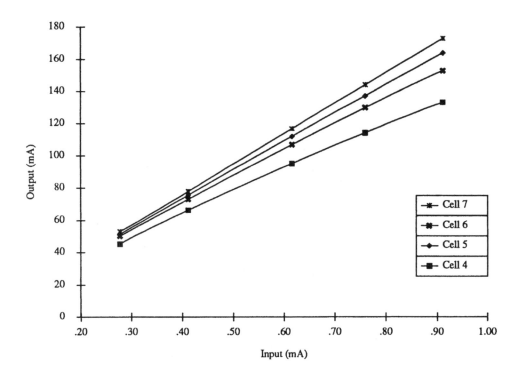

Figure 8.18   Intensity dependence of each cell, showing reflectance-intensity interdependence.

A preferentially weighted interconnection profile will produce an asymmetric receptive field[4] as shown in Fig. 8.21. A point source response thus is sufficient in establishing the preferred direction of response.

Although the above experiments were conducted with a one-dimensional array, the results can be extended to two-dimensional arrays to produce a direction preference of virtually any angle and localized to any area of the visual field. In such a case, it is desirable that the directional selectivity be dynamically programmable rather than pre-wired. This property is a subset of complete programmability of the connection strengths but is much simpler to implement since one-line per preferred direction can transmit the required control signal.

Another parameter of interest in Figs. 8.19 and 8.20 is the effect of the extent of connections on the network properties. It is seen from these figures that a network with

---

[4]Often the interconnection matrix is taken as the receptive field, which is not strictly correct, even though there exists strong correlation between the dendritic arborization and the shape and extent of the receptive field (Boycott and Dowling 1969).

Directional Selectivity

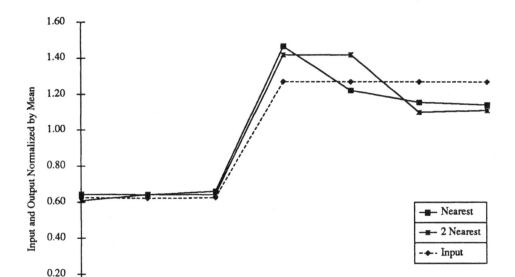

Figure 8.19   Directional selectivity (preferred direction); each cell receives
input from its left hand neighbors.

nearest neighbor connections is more responsive to an abrupt edge than one with two
nearest neighbor connections. Study of such correlations is planned for the future.

Finally, unidirectionality is an extreme case of asymmetry. Asymmetric networks
are hard to analyze and usually suffer from lack of stability. A case in point is an
asymmetric network, studied by Cohen and Grossberg (1983), which can be arbitrarily
close to a symmetric network but can be proven to be unstable. Unidirectional networks,
on the other hand, have lower (or upper) triangular connection matrices whose eigen
values are the diagonal terms and hence can be trivially examined for stability. The
stability problem was avoided in the study of preferential selectivity by choosing the
extreme case of unidirectional connections.

8.9 CONCLUSIONS

The general design considerations investigated in Chapters 6 and 7 were employed, and
thus verified, in the implementation of the simplest shunting recurrent network that can be
implemented within the design framework of Chapter 6. This implementation exhibited

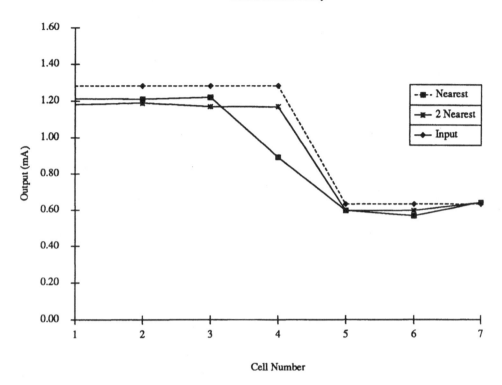

Figure 8.20 Directional selectivity (null direction); orientation of the input produces difference in response.

many of the predicted properties of the shunting networks which have been used to explain a variety of visual phenomena as mentioned in Chapter 3. The implementation also proved the suitability of the design for technological applications, specifically it was shown that:

- The network is capable of adapting to mean intensity level; the dynamic range of the output is compressed without negatively affecting the processing capability of the network. Based on circumstantial evidence it was hypothesized that the network is indeed capable of performing data compression, but no theoretical proof was presented

- The point source response of the network was seen to resemble the on-center off-surround receptive fields which are observed in many excitable cells and possess well known computational properties.

- It was shown that the receptive field, and in fact network response, varies with the variation of the mean intensity level in a desirable manner, that is, the network uses the available ambient light to better process the input data. This capability

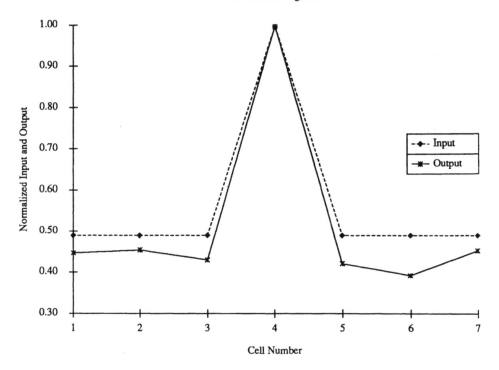

Figure 8.21   Receptive field of a unidirectional network wired for preferential directional and motion selectivity.

translates into the tuning of the modulation transfer function in the spatial frequency domain.

- The degree of the sensitivity of the network to mean intensity variation was shown to be externally controllable, which means that the network has the potential to *learn* from its operational environment and adapt its behavior accordingly. This tuning capability is in addition to the adaptation to mean intensity level that arises from multiplicative interaction. It was predicted that the same mechanism is biologically plausible.

- The network was shown to expand spatial edges and areas of high contrast, and the well-known psychophysical phenomenon of Mach bands was also observed. The amount of edge expansion was seen to be a function of mean input intensity and the cells with higher activity were shown to regulate their activity so as to preserve their dynamic range of response.

- Interaction of reflectance processing and intensity was demonstrated. An interpretation was provided by noting that the network has coded a "clue" to intensity while preserving its dynamic range of sensitivity.

- Preferential directional sensitivity was demonstrated by a unidirectionally wired network; the response was drastically different for a spatial edge in the preferred direction than the other. The resultant asymmetric receptive field also indicated the preferred direction.

Each of these properties are important and desirable, but the main distinguishing feature is that such a compact implementation exhibits *all* of these properties.

9

Nonlinear Mathematical Description

No study of a physical system is complete without a mathematical description, for, in Galileo Galilei's (1564–1642) words:

> The book of nature is written in the language of mathematics.

In this chapter a theoretical study of the network is performed which specifies important global properties of the network. Nonlinear system synthesis and identification methods of Volterra-Wiener series expansion is used in the following sections to better predict the system behavior and demonstrate the computational power embedded in multiplicative lateral inhibitory interaction.

9.1 STABILITY AND CONTENT ADDRESSABILITY

The emergent global properties of a network, rather than the behavior of the individual units and the local computation performed by them, describe the network's behavior.[1] Global stability analysis techniques, such as Liapunov energy functions, show the conditions under which a system approaches an equilibrium point in response to an input pattern. The equilibrium point is then the stored representation of the input. This property is termed Content-Addressable Memory property (CAM).

---

[1] Global computation has sometimes been taken (Blake and Zisserman 1987) to mean *cooperativity*. In the present work cooperation has been used in contradistinction to competition, both of which are facets of the global behavior.

Local stability, by contrast, involves the analysis of network behavior around individual equilibrium points.

The stability analysis is greatly facilitated by a theorem by Cohen and Grossberg (1983) who showed that a system of the general form,

$$\frac{dx_i}{dt} = a_i(x_i)\left[b_i(x_i) - \sum_{j=1}^{n} c_{ij}d_j(x_j)\right], \tag{9.1}$$

which includes many of the most popular neural network models as shown in Grossberg (1988a), admits the global Liapunov function,

$$V(x) = -\sum_{i=1}^{n}\int_0^{x_i} b_i(\xi_i)d_i'(\xi_i)d\xi_i + \frac{1}{2}\sum_{j,k=1}^{n} c_{jk}d_j(x_j)d_k(x_k), \tag{9.2}$$

under the constraints,

$$c_{ij} = c_{ji}, \tag{9.3}$$

$$a_i(x_i) \geq 0, \tag{9.4}$$

$$d_j'(x_j) \geq 0. \tag{9.5}$$

If the following substitutions are made in (9.1)

$$c_{ij} = k_{ij}, \tag{9.6}$$

$$d_j(x_j) = f_j(x_j) \quad \text{with} \quad f_j'(x_j) \geq 0, \tag{9.7}$$

$$b_i(x_i) = \left(\frac{I_i}{x_i} - \frac{1}{R_i}\right), \tag{9.8}$$

$$a_i(x_i) = \frac{x_i}{C_i} \quad \text{with} \quad x_i \geq 0, \tag{9.9}$$

which satisfy the symmetry, positivity, and monotonicity requirements as expressed in (9.3)–(9.5), then equation (9.1) is transformed into,

$$\frac{dx_i}{dt} = \frac{x_i}{C_i}\left[\left(\frac{I_i}{x_i} - \frac{1}{R_i}\right) - \sum_{j=1}^{n} k_{ij}f_j(x_j)\right], \tag{9.10}$$

and can be rewritten as,

$$C_i\frac{dx_i}{dt} = I_i - \frac{x_i}{R} - x_i\left[\sum_{j\neq i} k_{ij}f_j(x_j)\right]. \tag{9.11}$$

This is the original multiplicative lateral inhibitory equation, (3.1) of Chapter 3, and is thus shown to be globally asymptotically stable.

The global Liapunov function for (9.11) can be found by substituting (9.6)–(9.9) in (9.2) to obtain,

$$V(x) = -\sum_{i=1}^{n} \int_{0}^{x_i} \left(\frac{I_i}{\xi_i} - \frac{1}{R_i} \right) f_i'(\xi_i) d\xi_i + \frac{1}{2} \sum_{j,k=1}^{n} k_{jk} f_j(x_j) f_k(x_k). \qquad (9.12)$$

Since $f_j(x_j)$ is a nonnegative monotone nondecreasing function of x_j according to the constraint (9.7), the second term under the integral sign can be easily integrated and assuming, without loss of generality, that $f_j(0) = 0$, the Liapunov function is rewritten as,

$$V(x) = -\sum_{i=1}^{n} \int_{0}^{x_i} \frac{I_i}{\xi_i} f_i'(\xi_i) d\xi_i + \sum_{i=1}^{n} \frac{1}{R_i} f_i(x_i) + \frac{1}{2} \sum_{j,k=1}^{n} k_{jk} f_j(x_j) f_k(x_k). \qquad (9.13)$$

A problem is immediately apparent with this function. A constant term in $f_j'(\xi_i)$ will produce $\ln(\xi_i)$ when integrated, which equals $-\infty$ when the lower bound of the integral, zero, is substituted for ξ_i.

This problem can be solved by replacing the lower bound of the integral by a positive constant λ_i. It should then be shown that the new function,

$$V(x) = -\sum_{i=1}^{n} \int_{\lambda_i}^{x_i} \frac{I_i}{\xi_i} f_i'(\xi_i) d\xi_i + \sum_{i=1}^{n} \frac{1}{R_i} f_i(x_i) + \frac{1}{2} \sum_{j,k=1}^{n} k_{jk} f_j(x_j) f_k(x_k), \qquad (9.14)$$

has the Liapunov function property, that is, it satisfies

$$\frac{d}{dt} V(x(t)) \le 0 \qquad (9.15)$$

on all admissible trajectories.

This proposition is verified by noting that by chain rule

$$\frac{d}{dt} V(x(t)) = \sum_{i} \frac{\partial V(x(t))}{\partial x_i(t)} \frac{dx_i(t)}{dt}. \qquad (9.16)$$

Directly differentiating $V(x)$ with respect to x_i will result in

$$\frac{\partial V(x(t))}{\partial x_i(t)} = -\frac{\partial}{\partial x_i} \left(\int_{\lambda_i}^{x_i} \frac{I_i}{\xi_i} f_i'(\xi_i) d\xi_i \right) + \frac{f_i'(x_i)}{R_i}$$

$$+ \frac{1}{2} f_i'(x_i) \sum_{j} k_{ij} f_j(x_j)$$

$$+ \frac{1}{2} f_i'(x_i) \sum_{k} k_{ki} f_k(x_k). \qquad (9.17)$$

Using the symmetricity requirement (9.3) of $k_{ij} = k_{ji}$, and letting

$$g_i'(\xi_i) = \frac{I_i}{\xi_i} f_i'(\xi_i), \qquad (9.18)$$

(9.17) can be simplified to

$$\frac{\partial V(x(t))}{\partial x_i(t)} = -\frac{\partial}{\partial x_i}[g_i(x_i) - g_i(\lambda_i)] + \frac{f_i'(x_i)}{R_i} + f_i'\sum_j k_{ij}f_j(x_j). \tag{9.19}$$

Since $\partial g_i(\lambda_i)/\partial x_i = 0$, (9.19) can be substituted in (9.16), with dx_i/dt replaced by its definition (9.10), to yield

$$\frac{d}{dt}V(x(t)) = -\sum_{i=1}^n \frac{x_i f_i'(x_i)}{C_i}\left[\frac{I_i}{\xi_i} - \frac{1}{R_i} - \sum_{j=1}^n k_{ij}f_j(x_j)\right]^2. \tag{9.20}$$

Using constraints (9.7) and (9.9) which specify admissible trajectories, it is seen that

$$\frac{d}{dt}V(x(t)) \le 0, \tag{9.21}$$

which is the desired result.

It should be noted that the above derivation is basically a plausibility argument and for that reason has not been treated as a proof to a proposition. The proof can be completed by showing stability of *all* the trajectories. A rigorous proof that the integral $\int_0^{x_i}$ can be replaced by the an integral $\int_{\lambda_i}^{x_i}$ for the general case where

$$\lim_{\xi \to 0+} |b_i(\xi)d_i'(\xi)| = \infty \tag{9.22}$$

in (9.1) has been shown in Cohen and Grossberg (1983).

9.1.1 Stability of the Implemented Network

It was shown in Chapter 8 that the implemented network is accurately described by the network equation, (8.1), which is repeated here for ease of reference

$$\frac{dx_i}{dt} = I_i - a_i x_i + K_i x_i^2 - x_i \sum_{j \ne i} f_j(x_j). \tag{9.23}$$

This equation can be written in the general form of (9.1) by making the following change of variables,

$$c_{ij} = k_{ij}, \tag{9.24}$$

$$d_j(x_j) = f_j(x_j) \quad \text{with} \quad f_j'(x_j) \ge 0, \tag{9.25}$$

$$b_i(x_i) = \left(\frac{I_i}{x_i} - a_i + K_i x_i\right) \tag{9.26}$$

$$a_i(x_i) = x_i \quad \text{with} \quad x_i \ge 0, \tag{9.27}$$

which also show the network's compliance with the constraints and hypotheses that govern (9.1). Specifically, both of the circuits which were suggested for the implementation of the sigmoidal nonlinearity function $f_j(x_j)$ are monotone nondecreasing and smooth functions which satisfy (9.25).

The Liapunov energy function for the implemented network can be found by simply substituting (9.24)–(9.27) in (9.2) to get

$$V(x) = -\sum_{i=1}^{n} \int_{0}^{x_i} \left(\frac{I_i}{\xi_i} - a_i + K_i \xi_i \right) f_i'(\xi_i) d\xi_i + \frac{1}{2} \sum_{j,k=1}^{n} k_{ij} f_j(x_j) f_k(x_k). \quad (9.28)$$

For the special case of $k_{ij} f_j(x_j) = K_{ij} x_j$, which is the network the results of whose implementation were reported Chapter 8, an explicit Liapunov function can be derived,

$$V(x) = -\sum_{i=1}^{n} I_i \ln(\frac{x_i}{\lambda_i}) + \frac{a_i K_{ij}}{2} \sum_{i=1}^{n} x_i^2 - \frac{K_i K_{ij}}{3} \sum_{i=1}^{n} x_i^3 + \sum_{j,k=1}^{n} K_{jk} x_j x_k, \quad (9.29)$$

where λ_i is chosen according to the arguments of the previous section.

A crucial feature of the implemented network as well as the general Cohen and Grossberg model (9.1) is the extent of the excitatory center versus the inhibitory surround, or, in other words, the extent of cooperation versus competition. In Cohen and Grossberg (1983), based on numerical analyses and computer simulations, it was suggested that for the networks of the form,

$$\frac{dx_i}{dt} = -A_i x_i + (B_i - C_i x_i) \left[I_i + \sum_{k=1}^{n} D_{ik} f_k(x_k) \right]$$

$$- (E_i + F_i x_i) \left[J_i + \sum_{k=1}^{n} G_{ik} g_k(x_k) \right], \quad (9.30)$$

which generalize (9.1) by allowing extended excitatory interaction, an absolute stability result should exist. Counter-examples were later given (Cohen 1988) where subsets of (9.30) would consistently oscillate. As a general rule the excitatory center should be much narrower than the inhibitory surround, but stability of a network having wider than center-self excitatory interaction should be studied for each individual case.

This conclusion applies to the general network equation that can be implemented by the methods described in this work, namely (6.4), repeated here,

$$\frac{dx_i}{dt} = \pm I_i \pm a_i x_i \pm x_i(K_i x_i) \pm x_i \sum_{j \neq i} K_{ij} x_j \pm \sum_{j} K_j x_j^2. \quad (9.31)$$

While in (9.31) provision for arbitrary extent excitatory interaction is made, not all the possible configurations are stable and each case must be analyzed individually.

It is noteworthy that it was implicitly assumed here that memory is the pattern of the spatial activity of the cells of a neural network *in steady state*. It was for this reason that the network equations were called short-term memory equations and the stability of the steady state was treated with special significance and termed content addressable memory. Oscillations, bursts, travelling waves, and even chaotic behavior have also been suggested as likely mechanisms of (short-term) memory and there is on-going debate on the validity of each hypothesis with respect to the rest. For present purposes stable steady state behavior as short term memory has been accepted as a working hypothesis.

The proof of the CAM property extends the use of the network from sensory systems to any system which requires short term memory capability. Specifically, in networks which learn by changing some internal parameters, the proof of CAM property means that irrespective of the inputs, changing parameters, or initial values of the model, the network remains stable in response to an unpredictable environment. The network is thus compatible with higher order processing. This should not be surprising since in Chapter 1 it was mentioned that shunting networks have also been reported to exist in several areas of the brain.

There exists a neural network architecture, the Adaptive Resonance Theory (ART), which employs shunting networks coupled with long term memory equations, in conjunction with other controlling mechanisms, to achieve stable *learning*. The relation of the implemented network with that architecture will be investigated next.

9.2 RELATION TO ADAPTIVE RESONANCE

Adaptive resonance architectures are neural networks that produce stable recognition codes in response to arbitrarily complex sequences of input patterns and do so by self-organizing, i.e., without requiring an external teacher. The principles of ART were introduced by Grossberg (1976) and have since been used to qualitatively analyze data about, among others, speech perception, word recognition and recall, visual perception, olfactory coding, evoked potentials, thalmacortical interactions, and amnesia (Carpenter and Grossberg 1987a). Two general classes of adaptive resonance *architectures*, ART1 and ART2, have been introduced by Carpenter and Grossberg (1987a, b) and were shown to be completely described by sets of differential equations.

A schematic diagram of the different modules of a typical ART architecture is shown in Fig. 9.1. The operation of an ART circuit is, briefly,[2] as follows. An input pattern is presented to the feature representation field, F1, of an adaptive resonance circuit. The input is contrast-enhanced and transformed into a pattern of Short Term Memory (STM) activity across F1. This pattern of activity is multiplied, or *gated*, by adaptive Long Term Memory (LTM) traces and then presented to the pattern representation field, F2. The F2 level is a competitive network that produces an output which is the "code" for the input pattern. Usually the F2 node which is the most active suppresses all the other nodes of F2. This is the case of a "winner-take-all" circuit where a choice has been made as the extreme of competition. The network *learns* the input by storing its

---

[2]Detail of the operation of ART, its application from psychology to sensory-motor control, and its comparison with other learning models can be found in Grossberg (1976, 1988a), and Carpenter and Grossberg (1987a, b, 1988). An introductory tutotrial appears in Nabet (1988).

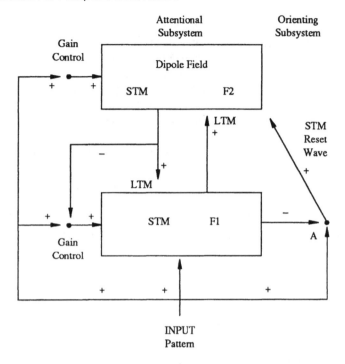

Figure 9.1 A modulatory view of adaptive resonace theory. LTM and STM
are short term memory and long term memory, respectively, ρ is the vigilance
parameter, and positive and negative signs show excitatory and inhibitory interaction respectively. Adopted from Carpenter and Grossberg (1987a).

code, or representation, in long term memory via a formula, such as

$$\frac{d}{dt}z_{ij} = \epsilon x_j(-z_{ij} + x_i), \tag{9.32}$$

where z_{ij} is the long term memory trace that gates node x_i of F1 to x_j of F2, and ϵ is a
scaling parameter. The first term of the learning rule ensures exponential decay in lieu
of active input and the second term signifies the fact that the *synaptic efficacies* change
according to the *product* of input and output. This is the famous Hebbian learning rule
after Hebb (1949).

The system described so far is a competitive learning system which, although
very powerful, can become unstable in response to complex environment. That is, the
same input pattern may produce a different recognition code depending on the history of
learning, or oscillate between two or more representations without settling on any.

The rest of the circuitry of Fig. 9.1 ensures the stability of learning by using top-
down learned expectancies and engaging in adaptation of the long term memory traces,
learning, only when the expectation and the input mutually reinforce each other in a
resonance state; hence the name "adaptive resonance." A vigilance parameter ρ sets the

learning resolution, that is, the degree to which two patterns should differ before enough novelty is detected to warrant a new categorization.

Adaptive resonance theory, it should be clarified, is a *theory*, a computational construct. ART1 and ART2 are *circuits* which realize, or implement, that theory. The important feature of both ART1 and ART2 is that they isolate mechanisms of adaptive resonance and provide differential equations which describe these modules, with ART1 accepting only binary input patterns and ART2 working with both analog and binary inputs.

The feature representation field F1 of both ART1 or ART2 circuits is described by

$$\epsilon \frac{d}{dt} x_i = -A x_i + (1 - B x_i) J_i^+ - (C + D x_i) J_i^-, \qquad (9.33)$$

where the term J_i^+ is the total excitatory input to the ith node and J_i^- is the total inhibitory input, and ϵ represents the ratio between the STM relaxation time and LTM relaxation time.

For the special case where $A = a_i$, $B = 0$, $C = 0$, $D = 1$, and,

$$J_i^- = \sum_{i \neq j} K_{ij} x_j, \qquad (9.34)$$

$$J_i^+ = I_i + K_i x_j^2, \qquad (9.35)$$

equation (9.33) reduces to

$$\frac{d}{dt} x_i = I_i - a_i x_i + K_i x_i^2 - x_i \sum_{i \neq j} K_{ij} x_j, \qquad (9.36)$$

which is the implemented shunting network. Note that ϵ is irrelevant here since there is no long term memory trace.

This result shows that the implemented network can readily function as a front-end processor for an adaptive resonance circuit. But in order to be completely compatible with an ART architecture, the implemented network needs to accommodate variable K_{ii}s to allow for gating by long term memory traces of the top-down expectancies. As mentioned previously a design with variable connection strengths is under fabrication and will be reported elsewhere.

The F2 field is also a (shunting) competitive network and can be implemented by the designs introduced here. F2 fields should also allow for LTM gating and furthermore it should be possible to selectively deactivate, or reset, the active F2 nodes during a cycle of the learning process. Designs that include both provisions have been studied but a serious hardware difficulty arises from the large extent of required connectivity of an ART architecture.

9.3 VOLTERRA-WIENER SERIES EXPANSION

Linear mathematical methods of system identification are incapable of satisfactorily identifying nonlinear systems.

The nonlinear analysis technique of Volterra-Wiener expansions has been success-fully applied to, among others, the analysis of wave propagation in random media, scattering by random surfaces, Brownian motion, theory of turbulence, and biological systems (for references see Ogura 1986). In the next section this method is applied to the analysis of temporal and spatial characteristics of neural networks.

9.4 TEMPORAL KERNELS

The objective is to write the solution to the set of differential equations

$$\frac{dx_i}{dt} = I_i - a_i x_i + K_i x_i^2 - x_i \left(\sum_j K_{ij} x_j \right), \tag{9.37}$$

which accurately characterize the implemented network, in the Wiener-Volterra integral expansion form,

$$x_i = H_{i0} + \int_{-\infty}^{\infty} \mathbf{H}_{i1}(\tau) \cdot \mathbf{l}(t - \tau) d\tau$$

$$+ \int_{-\infty}^{\infty} \int_{-\infty}^{\infty} \mathbf{H}_{i2}(\tau_1, \tau_2) \cdot \mathbf{l}(t - \tau_1) \otimes \mathbf{l}(t - \tau_2) d\tau_1 d\tau_2 + \cdots \tag{9.38}$$

In (9.38), $\mathbf{H}_{ij}(\tau_1, \tau_2, \ldots, \tau_j)$ is the ith row of the jth kernel and $\otimes$ is the *Kronecker product* used to denote multi-dimensional dot products, and is defined by

$$A \otimes B = \begin{bmatrix} a_{11}B & \cdots & a_{1m_a}B \\ \vdots & & \vdots \\ a_{n_a1}B & & a_{n_am_a}B \end{bmatrix}$$

for matrices $A = [a_{ij}]$ and $B = [b_{ij}]$, of dimension $n_a \times m_a$ and $n_b \times m_b$, respectively.

The first integral of (9.38), is the familiar convolution operator and the higher order terms show the inclusion of higher order nonlinearities in analysis and identification of nonlinear systems. A linear analysis technique, by contrast, is capable of only identifying the first two terms of (9.38).

The existence of the convergent Volterra series is shown for a large class of sys-tems in Boyd and Chua (1985). Study of Wiener-Volterra systems with application to identification and synthesis of physiological systems is done in a text by Marmarelis and Marmarelis (1978). Schetzen (1980), and Rugh (1981) have provided classic treatments of Volterra-Wiener approach from a system theoretic point of view. The Wiener kernels can be found by the familiar Gram-Schmit orthogonalization procedure, the formulas for which are discussed in all of these texts.

A variational analysis technique (Rugh 1981, Pinter 1987a) can be applied to (9.37) to produce a solution in the desired form. The kernels can then be extracted by observation and the input-output relationship is modeled more accurately as compared to a linearized model. This technique which is also extensively used in quantum mechanical analysis (Yariv 1975) to find, for instance, higher order Hamiltonians of Schrödinger's equation, is outlined below.

1. Let the input $I_i = L_0 + \lambda l_i$ where λ is the perturbation parameter.

2. Assume a solution of the form $\mathbf{x} = \bar{\mathbf{x}} + \lambda \mathbf{x_1} + \lambda^2 \mathbf{x_2} + \cdots$, and differentiate it.

3. Insert this solution in the original differential equation.

4. Equate like powers of λ in equations which result from steps 2 and 3 above.

The following derivation proceeds according to this outline.
Let the input

$$I_i(t) = L_0 + \lambda l_i(t), \tag{9.39}$$

where L_0 is the input background, uniform across the array and independent of time, and λ is the perturbation parameter, be applied to the system of (9.37). The response of the network is

$$\frac{dx_i}{dt} = L_0 + \lambda l_i(t) - a_i x_i + K_i x_i^2 - x_i \left(\sum_j K_{ij} x_j \right). \tag{9.40}$$

The *total* input is then the vector $\mathbf{I} = [L_0 + l_1, L_0 + l_2, \ldots, L_0 + l_n]^t$, and the output vector is $\mathbf{x} = [x_1, x_2, \ldots, x_n]^t$.

Next, the standard assumption is made that the response to the perturbed input can be written in the form

$$x_i = \lambda^0 \bar{x}_i + \lambda^1 x_{i1} + \lambda^2 x_{i2} + \lambda^3 x_{i3} + \cdots, \tag{9.41}$$

where x_i is the ith term of the output vector and the second index of the subscript notates the order of the term which is the order of approximation.

Differentiating this equation leads to

$$\dot{x}_i = \lambda^0 \times 0 + \lambda^1 \dot{x}_{i1} + \lambda^2 \dot{x}_{i2} + \lambda^3 \dot{x}_{i3} + \cdots. \tag{9.42}$$

Substituting the assumed form of solution (9.41) in the perturbed differential equation (9.40) results in

$$
\begin{aligned}
\dot{x}_i = {} & L_0 + \lambda l_i - a_i [\lambda^0 \bar{x}_i + \lambda^1 x_{i1} + \lambda^2 x_{i2} + \lambda^3 x_{i3} + \cdots] \\
& + K_i [\lambda^0 \bar{x}_i + \lambda^1 x_{i1} + \lambda^2 x_{i2} + \lambda^3 x_{i3} + \cdots]^2 \\
& - [\lambda^0 \bar{x}_i + \lambda^1 x_{i1} + \lambda^2 x_{i2} + \lambda^3 x_{i3} + \cdots] \\
& \quad \sum_j K_{ij} [\lambda^0 \bar{x}_j + \lambda^1 x_{j1} + \lambda^2 x_{j2} + \cdots].
\end{aligned}
\tag{9.43}
$$

Equations (9.42) and (9.43) should hold for all values of λ, which means that like powers of λ should have the same coefficients. Equating the terms that are multiplied by λ^0 in (9.42) and (9.43) results in

$$L_0 - a_i\bar{x}_i + K_i\bar{x}_i^2 - \bar{x}_i\sum_j K_{ij}\bar{x}_j = 0, \tag{9.44}$$

which is the steady state response of the network to the uniform input, L_0.

One solution to (9.44) is $\bar{x}_i = \bar{x}_j$ for all j, that is, the response of a symmetrically connected network to a uniform input is uniform across the array. This was previously argued to be the case when response to uniform input was taken to be the same as the response of one cell with self inhibition to a single input (Fig. 8.1), and was also experimentally verified in Fig. 8.3. This transforms (9.44) into

$$\bar{x}_i^2[K_i - \sum_j K_{ij}] - a_i\bar{x}_i + L_0 = 0 \quad \text{for all } i. \tag{9.45}$$

Noting that K_i is the quadratic term that arises from the implementation and is defined as

$$K_i \equiv \frac{1}{2}\sum_{j\neq i}^n K_{ij}, \tag{9.46}$$

(9.45) further simplifies to the quadratic algebraic equation

$$-K_i\bar{x}_i^2 - a_i\bar{x}_i + L_0 = 0, \tag{9.47}$$

which has the solution

$$\bar{x}_i = \frac{-a_i + \sqrt{a_i^2 + 4L_0K_i}}{2K_i}, \tag{9.48}$$

where only the solution which satisfies the positivity constraint $\bar{x}_i \geq 0$ has been chosen. This response to background intensity is the zero order kernel, namely,

$$H_{0i} = \bar{x}_i. \tag{9.49}$$

Equation (9.48) clearly demonstrates that the output is proportional to the square root of the uniform input. This is the dynamic range compression which was extensively described in Chapter 8.

From this treatment it is seen that in general if the order of $x_i f_i(x_i)$ is n, that is, if

$$x_i f_i(x_i) \doteq \alpha x_i^n + \beta x_i^{n-1} + \cdots + \gamma, \tag{9.50}$$

then the input beyond a lower bound is compressed by an order of $1/n$, or

$$\bar{x}_i \propto [L_0]^{\frac{1}{n}} \tag{9.51}$$

Specifically, if the feedback nonlinearity is exponential within a range, the output will be approximately proportional to the logarithm of the input within that range.

9.4.1 First Order Temporal Kernel

The first order Volterra kernel which performs the familiar (linear) convolution operation is found by equating the coefficients of λ^1 to obtain

$$\dot{x}_{i1} = l_i(t) + x_{i1}\left(-a_i + \bar{x}_i 2K_i - \sum_j K_{ij}\bar{x}_j\right) - \bar{x}_i \sum_j K_{ij}x_{j1}. \tag{9.52}$$

For simplicity let

$$b_i = a_i - \bar{x}_i 2K_i + \sum_j K_{ij}\bar{x}_j, \tag{9.53}$$

which reduces to $b_i = a_i$ when identity (9.46) is used. In matrix notation (9.52) can be written as

$$\begin{bmatrix} \dot{x}_{11} \\ \dot{x}_{21} \\ \vdots \\ \dot{x}_{n1} \end{bmatrix} = -\begin{bmatrix} b_1 & \bar{x}_1 K_{12} & \bar{x}_1 K_{13} & \dots & \bar{x}_1 K_{1n} \\ \bar{x}_2 K_{21} & b_2 & \bar{x}_2 K_{23} & \dots & \bar{x}_2 K_{2n} \\ \vdots & & \ddots & & \vdots \\ \bar{x}_n K_{n1} & \dots & & & b_n \end{bmatrix}\begin{bmatrix} x_{11} \\ x_{21} \\ \vdots \\ x_{n1} \end{bmatrix} + \begin{bmatrix} l_1 \\ l_2 \\ \vdots \\ l_n \end{bmatrix} \tag{9.54}$$

and can further be simplified by noting that $\bar{x}_i = \bar{x}_j$, but is left in the more general form to allow further refinement by, for example, using linear programming techniques to analyze the zero order kernel.

Equation (9.54) is in the form

$$\dot{\mathbf{x}}_1 = \mathbf{A}\mathbf{x}_1 + \mathbf{l} \tag{9.55}$$

and can be solved for $\mathbf{x}$ by using techniques of linear algebra. It is important to note that matrix $\mathbf{A}$ is dependent on the 0th order kernel which is in turn dependent on the mean intensity of the input. The Volterra expansion technique is thus capable of propagating the intensity dependence of the model to higher order kernels.

The solution to (9.55) is

$$\mathbf{x}_1(t) = e^{\mathbf{A}(t-t_0)}\mathbf{x}_1(t_0) + \int_{t_0}^{t} e^{\mathbf{A}(\tau)}\mathbf{l}(t-\tau)d\tau. \tag{9.56}$$

Assuming, without loss of generality, a time origin of zero and zero initial conditions, that is, $t_0 = 0$ and $\mathbf{x}_1(t_0) = \mathbf{0}$, the generating solution (9.56) can be written as

$$\mathbf{x}_1(t) = \int_{t_0}^{t} e^{\mathbf{A}(\tau)}\mathbf{l}(t-\tau)d\tau. \tag{9.57}$$

The state transition matrix $e^{\mathbf{A}(t)}$ can be found by taking the inverse Laplace transform of $(S\mathbf{I} - \mathbf{A})^{-1}$, that is,

$$e^{\mathbf{A}(t)} = \mathcal{L}^{-1}(S\mathbf{I} - \mathbf{A})^{-1}. \tag{9.58}$$

The roots of the characteristic function $|S\mathbf{I} - \mathbf{A}| = 0$ are the eigenvalues and can, in general, be repeated or nondistinct, but since the coefficient matrix is symmetric ($K_{ij} = K_{ji}$), from (9.54) it is observed that $\mathbf{A}$ is Hermitian and hence has real eigenvalues and can be diagonalized even if the eigenvalues are not distinct (Franklin 1968). Proof of global asymptotic stability in the previous section is also equivalent to proving that the eigenvalues of $\mathbf{A}$ are negative. From these considerations the state transition matrix can be written as

$$e^{\mathbf{A}(t)} = \begin{bmatrix} \sum_{k=1}^{n} a_{11k}e^{a_k t} & \sum_{k=1}^{n} a_{12k}e^{a_k t} & \cdots & \sum_{k=1}^{n} a_{1nk}e^{a_k t} \\ \vdots & \vdots & & \vdots \\ \vdots & \vdots & & \vdots \\ \sum_{k=1}^{n} a_{n1k}e^{a_k t} & \sum_{k=1}^{n} a_{n2k}e^{a_k t} & \cdots & \sum_{k=1}^{n} a_{nnk}e^{a_k t} \end{bmatrix}, \tag{9.59}$$

where the eigenvalues a_k are negative. Had the $\mathbf{A}$ matrix not been Hermitian, possible nondistinct eigenvalues would produce terms of the form $\sum_k b_k t^k e^{a_k}$ which would considerably complicate the derivation. Furthermore, the proof of stability was necessary in order to argue that $\mathbf{A}$ is negative definite.

Finding the state transition matrix in general is not a trivial problem and often numerical methods should be employed for its determination. The above derivation shows the existence of a closed form solution, the general shape of that matrix, and specifically its relevance to Volterra series expansion.

The first order generating solution is

$$\begin{bmatrix} x_{11} \\ x_{21} \\ \vdots \\ x_{n1} \end{bmatrix} = \begin{bmatrix} \sum_{j=1}^{n} \int_0^t \sum_{k=1}^{n} a_{1jk}e^{a_k \tau} l_j(t - \tau)d\tau \\ \vdots \\ \vdots \\ \sum_{j=1}^{n} \int_0^t \sum_{k=1}^{n} a_{njk}e^{a_k \tau} l_j(t - \tau)d\tau \end{bmatrix}. \tag{9.60}$$

If (9.41) is rewritten, by observing that the perturbation parameter may be implicit as a scaling parameter in the kernels, as

$$\mathbf{x} = \bar{\mathbf{x}} + \mathbf{x}_1 + \mathbf{x}_2 + \cdots, \tag{9.61}$$

then

$$\mathbf{x} = \bar{\mathbf{x}} + \int_0^t e^{\mathbf{A}\tau} \mathbf{l}(t - \tau)d\tau + \mathbf{x}_2 + \cdots, \tag{9.62}$$

it is observed, by comparing (9.62) and (9.38), that $e^{\mathbf{A}\tau}$ is indeed the desired first order kernel which, as previously mentioned, is dependent upon the 0th kernel and ultimately background intensity. Hence

$$\mathbf{H}_1(\tau) = e^{\mathbf{A}\tau} \tag{9.63}$$

where each term of the Volterra kernel is

$$\mathbf{H}_{i1}(\tau) = \sum_k a_{ijk} e^{a_k \tau}, \qquad j = 1, 2, \ldots \tag{9.64}$$

9.4.2 Second Order Temporal Kernel

The second order kernel is found by equating coefficients of λ^2 to get

$$\dot{x}_{i2} = x_{i2}\left(-a_i + \bar{x}_i 2K_i - \sum_j K_{ij}\bar{x}_j\right) - \bar{x}_i\sum_j K_{ij}x_{j2} + K_i x_{i1}^2 - x_{i1}\sum_{j\neq i} K_{ij}x_{j1} \tag{9.65}$$

and is seen to be of the same form as (9.52) but with the input l_i replaced by $K_i x_{i1}^2 - x_{i1}\sum_{j\neq i} K_{ij}x_{j1}$. Equation (9.65) can be written in the matrix form

$$\begin{bmatrix} \dot{x}_{12} \\ \dot{x}_{22} \\ \vdots \\ \dot{x}_{n2} \end{bmatrix} = \mathbf{A}\begin{bmatrix} x_{12} \\ x_{22} \\ \vdots \\ x_{n2} \end{bmatrix} + \begin{bmatrix} m_1 \\ m_2 \\ \vdots \\ m_n \end{bmatrix}, \tag{9.66}$$

where

$$m_i = x_{i1}\sum_{j=1}^n K'_j x_{j1} \tag{9.67}$$

is a function of the first order solution and

$$K'_j = \begin{cases} -K_{ij} & \text{for } j \neq i \\ K_i & \text{for } j = i \end{cases}. \tag{9.68}$$

Again assuming that $\mathbf{x}(0) = 0$, the second order generating solution is

$$\mathbf{x}_2(t) = \int_{t_0}^t e^{\mathbf{A}(\tau)}\mathbf{m}(t-\tau)d\tau. \tag{9.69}$$

The ith term of $\mathbf{x}_2(t)$ is found from (9.59) to be

$$x_{i2}(t) = \int_0^t \sum_l \sum_k a_{ilk} e^{a_k \tau} m_l(\tau)d\tau. \tag{9.70}$$

If $m_l(t)$ is replaced by its definition (9.67) with the first order solution explicitly substituted from (9.60), this equation becomes

$$x_{i2}(t) = \int_0^t \sum_l \sum_k a_{ilk} e^{a_k \tau}\left[\sum_j \int_0^{t-\tau}\sum_m a_{ljm}e^{a_m \tau_1}l_j(t-\tau-\tau_1)\right]d\tau_1$$

$$\left[\sum_p K'_p \sum_q \int_0^{t-\tau}\sum_r a_{pqr}e^{a_r \tau_2}l_q(t-\tau-\tau_2)\right]d\tau_2 d\tau. \tag{9.71}$$

In order to put this equation in a form from which the second order Volterra kernel can be extracted by observation, the following change of variables should be made

$$\mu = \tau + \tau_1$$

$$\gamma = \tau + \tau_2,$$

and for simplicity of notation let

$$c_i = \sum_k \sum_l \sum_m \sum_p \sum_r a_{ilk} a_{ljm} a_{pqr} K'_p. \tag{9.72}$$

Making these substitutions in (9.71) while carefully treating the limits of the integrals results in

$$x_{i2}(t) = \sum_j \sum_q c_i \int_0^t \int_0^t \left[\int_0^t e^{\tau(a_k - a_m - a_r)} e^{a_m \mu} e^{a_r \gamma} u(\mu - \tau) u(\gamma - \tau) d\tau \right]$$

$$l_j(t - \mu) l_q(t - \gamma) d\mu d\gamma \tag{9.73}$$

where $u(\alpha)$ is the step function defined in the usual manner.

Comparison of (9.73) and (9.38) shows that the quantity in brackets describes the terms of the second order temporal Volterra kernel and can be integrated to produce

$$h_{jq}(\tau_1, \tau_2) = \sum_m \sum_r \sum_k c'_i \frac{e^{a_m \tau_1} e^{a_r \tau_2}}{(a_k - a_m - a_r)} \left[e^{(a_k - a_m - a_r) \min(\tau_1, \tau_2)} - 1 \right], \tag{9.74}$$

where

$$c'_i = \sum_l \sum_p a_{ilk} a_{ljm} a_{pqr} K'_p. \tag{9.75}$$

This is the same form as that derived by Pinter (1987a) even though the model defined by Pinter does not have the quadratic self excitatory term that arises from the implementation. The reason for similarity lies in the fact that equations with the self-excitatory term are still globally asymptotically stable as proven in the previous chapter and the A matrix remains Hermitian. These were the constraints that led to the general form of the solution (9.60) hence the final result should be similar in form although the eigenvalues are different in the two cases.

It is seen that the second order temporal kernel consists of permutations of the products of exponential terms with eigenvalues and eigenfunctions of the A matrix as rate constants. In both the first and the second order temporal kernels, Eq. (9.64) and (9.74), the eigenvalues become larger as the mean intensity, L_0, increases. Thus temporal behavior of the kernels becomes faster and moves closer to the time origin, and kernels become smaller in magnitude. This is the basic temporal adaptation (Rushtonian Transformation in vision science), and is propagated to all higher order kernels (Pinter and Nabet 1990).

It is noteworthy that in physical implementations, even in the solid state integrated circuit implementations which produce very uniform device characteristics, it is hard to maintain the absolute symmetry of the interconnections. This may seem to invalidate the symmetry constraint which was repeatedly used in this chapter, but a theorem by Franklin (1968) can be invoked which states that such a matrix may be arbitrarily close to a diagonalizable matrix and hence the results of this chapter hold with arbitrary accuracy for a physically implemented case.

9.5 SPATIAL KERNELS

The same technique as described above can be applied to the steady state solution of (9.37), the existence of which is due to the global stability of the network (Nabet et al. 1989, Cohen and Grossberg 1983), to find the solution in the spatial Wiener-Volterra expansion form:

$$x_i = H_{0i} + \sum_j H_1(i,j)l_j + \sum_j \sum_k H_2(i,j,k)l_j l_k + \cdots, \qquad (9.76)$$

where the kernel subscript denotes the order of the kernel, which is the order of approximation, and the kernels are written with spatial indices as argument to stress the spatial dependence.

Extraction of the spatial Volterra kernels, however, is simpler if the results of the temporal case are used, as shown below.

9.5.1 Zero Order Kernel

The 0th spatial kernel has already been derived as the response to the (uniform) background intensity in steady state and is shown in equation (9.48). This response, as expected, is invariant to time and across the space as indexed by the cell number, assuming boundary effects are compensated by, for example, imposing periodic boundary condition. As previously mentioned both the spatial uniformity of the response and its dependence on background intensity level, for the connectivity described here, have been experimentally verified in Chapter 8.

9.5.2 First Order Spatial Kernel

The first order spatial response is found by solving the first order network equation, (9.37), which is the linearized model of the network, in the steady state (Pinter 1987a). Thus,

$$\mathbf{x}_1 = -\mathbf{A}^{-1}(L_0)\mathbf{l}, \qquad (9.77)$$

where $\mathbf{A}$ and $\mathbf{l}$ are as defined previously and the dependence of matrix $\mathbf{A}$ on the mean intensity via the zero order kernel is explicitly shown. The ith term of the first order response is simply

$$x_{i1} = \sum_j b_{ij} l_j, \qquad (9.78)$$

from which the ith term of the first order kernel is observed to be

$$H_1(i,j) = b_{ij}, \qquad (9.79)$$

where b_{ij} are the terms of the negative of the inverted matrix $\mathbf{A}^{-1}$, the existence of which is a result of the fact that $\mathbf{A}$ is positive definite.

Contrary to the temporal kernel case a closed form analytical expression for $\mathbf{A}^{-1}$ can not be written.[3] For the case where $\mathbf{A}$ is a Toeplitz matrix, which is applicable here, a direct relationship between $\mathbf{A}$ and its inverse exists (Pinter 1989) but that relationship which involves a fourier transform is not directly related to present discussion.

From (9.77) it is seen that $\mathbf{A}^{-1}(L_0)$ specifies the receptive field of the network. A column of the inverted matrix is plotted in (Pinter and Nabet, 1990) for different values of the mean intensity L_0 and is reproduced in Figure 9.2. The on-center off-surround shape of the response and its dependence on the mean intensity level is very similar to the experimental results of Chapter 8 (Figs. 8.8–8.10). This shows the suitability of the Volterra-Wiener expansion method for analyzing the intensity dependence behavior of the network.

It should be emphasized that although the first order kernel is derived from a linearized system model, its intensity dependence is due to the multiplicative nonlinearity which gave rise to the intensity dependent zero order kernel. A comparable additive model would not produce intensity dependent kernels.

9.5.3 Second Order Spatial Kernels

The second order spatial Volterra kernel is found, in a similar manner to the first order kernel, by finding the steady state solution of (9.66), namely

$$\mathbf{x}_2 = -\mathbf{A}^{-1}\mathbf{m}, \tag{9.80}$$

where $\mathbf{m}$ is defined as in (9.67). The dependence of the higher order kernels on the mean intensity via the zero order kernel is apparent from this equation. Furthermore, it is notable that the matrix inversion is performed only once and the results are propagated through the higher order kernels.

Each term of $\mathbf{x}_2$ is

$$x_{i2} = \sum_j b_{ij} m_j, \tag{9.81}$$

where as before the second index of the response specifies the order of the kernel.

Substituting the definition of m_j from (9.67), inserting the expression for the first order response from (9.78), and reordering the terms results in

$$x_{i2} = \sum_j \sum_k \left[\sum_p \sum_q K_q' b_{ip} b_{pj} b_{qk} \right] l_j l_k. \tag{9.82}$$

The term within the brackets is the desired ith term of the second order spatial kernel, namely

$$H_2(i, j, k) = \sum_p \sum_q K_q' b_{ip} b_{pj} b_{qk}. \tag{9.83}$$

---

[3]Common matrix inversion methods such as the cofactor method are *algorithms* rather than analytical expressions.

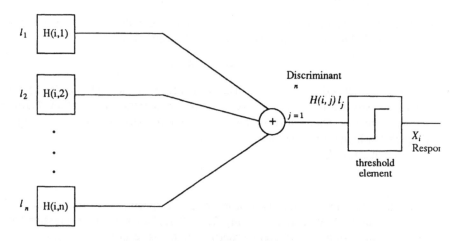

Figure 9.2 Receptive field of the linearized network plotted with different mean intensity values. From Pinter and Nabet (1990).

9.5.4 Higher Order Spatial Kernels

The same procedure can be applied to find any order kernel, assuming that in the steady state the derivative of each of the terms is finite, an assumption which is plausible due to the convergence of the series and asymptotic global stability of the network. In general the mth order spatial kernel is found by equating the coefficients of λ^m in (9.43) and observing that the mth order response is the steady state form of the resulting equation, namely,

$$x_{im} = \sum_{j=1}^{n} b_{ij} p_{jm},\qquad(9.84)$$

where, as before, b_{ij} are the terms of the negative of the inverted matrix **A** and p_{im} is a perturbation input term which depends on all the previous responses in the following manner

$$p_{im} = x_{i1}\sum_{j=1}^{n} K_{ij}x_{j,m-1} + x_{i2}\sum_{j=1}^{n} K_{ij}x_{j,m-2} + \cdots + x_{i,m-1}\sum_{j=1}^{n} K_{ij}x_{j1},\qquad(9.85)$$

where $K_{ii} = 0$ if the original multiplicative lateral inhibitory network, Eq. (3.2), is intended and $K_{ii} = -K_i$ if the implemented model of (9.37) is of interest.

This identity can be more compactly written as

$$p_{im} = \sum_{k=1}^{m-1}\sum_{j=1}^{n} K_{ij}x_{i,k}x_{j,m-k},\qquad(9.86)$$

which, from (9.84), leads to a steady state mth order spatial response of

$$x_{im} = \sum_{l=1}^{n}\sum_{k=1}^{m-1}\sum_{j=1}^{n} K_{ij}b_{il}x_{ik}x_{j,m-k}.\qquad(9.87)$$

Equation (9.87) is a recursive formula from which the mth order response can be derived based on the lower-order responses.

The third order response, for example, is found by simply letting $m = 3$ in (9.87) to obtain

$$x_{i3} = \sum_{l=1}^{n}\sum_{j=1}^{n} K_{ij}b_{il}x_{i1}x_{j2} + \sum_{l=1}^{n}\sum_{j=1}^{n} K_{ij}b_{il}x_{i2}x_{j1}.\qquad(9.88)$$

Substituting the expressions for first- and second-order responses (9.78) and (9.82), and reordering the terms results in

$$x_{i3} = \sum_{j}\sum_{p}\sum_{q}\left[\sum_{p2}\sum_{p3}\sum_{p4}\sum_{p5} K_{ij}K_{jp_3}b_{ip_2}b_{iq}b_{jp}b_{jp_4}b_{p_3,p_4}\right]l_j l_p l_q.\qquad(9.89)$$

It is observed that the term within the brackets is the third order kernel, namely

$$H_i(i,j,p,q) = \left[\sum_{p2}\sum_{p3}\sum_{p4}\sum_{p5} K_{ij}K_{jp_3}b_{ip_2}b_{iq}b_{jp}b_{jp_4}b_{p_3,p_4}\right].\qquad(9.90)$$

The recursive nature of (9.87) is especially suited for numeric calculation by a digital computer and is facilitated by the fact that the most computationally intensive operation of inversion of a matrix is done only once and recalled later. On the other hand recursion disallows writing of an independent closed form formula for the mth order kernel, the necessity for which is dependent on the rate of the convergence of the Volterra series. Since the significance of the contribution of kernels higher than the third order to the series convergence is unknown in many practical cases of interest, Eq. (9.87) is sufficient for present purposes.

The major aim of this treatment was to show that both the *one-layer* implemented network and the simplest network which consist of only multiplicative inhibitory and linear feedback terms can be represented as in (9.76). If this output is further transformed by placing a nonlinearity $S[\cdot]$ in front of each cell, such as a hard limiting circuit or a sigmoidal nonlinearity such as the ones suggested in Chapter 5, then the output is

$$x_i = S\left[H_0 + \sum_j H_1(i,j)l_j + \sum_j \sum_k H_2(i,j,k)l_j l_k + \cdots\right]. \qquad (9.91)$$

Each unit of the network thus not only calculates the weighted sum of the input, but is also sensitive to the higher order correlations as demonstrated by the fact that the third term of the above expression is the definition of a weighted (spatial) auto correlation function of the input. A unit defined by (9.91), has special classification properties which are discussed next.

9.5.5 Classification Properties

Typical units of (additive) neural network architectures transform a weighted sum of the input as in the classic Threshold Logic Units (TLU) of McCulloch and Pitts (1943), which were shown to be able to realize *any* logical function. A TLU is defined by

$$x_i = S\left[\sum_{j=1}^{n} H(i,j)l_j\right], \qquad (9.92)$$

where x_i is the output of the unit, S is a nonlinearity, l_j are the inputs, and $H(i,j)$ are weighting terms, also called connection strengths, written in this form to underline similarity with the Volterra kernel expansion.

The quantity inside the brackets is a *discriminant function* which specifies the *decision boundaries* for the classification of the input pattern (Nilsson 1965, Minsky 1967). If the nonlinearity S is a threshold element, as is the case for a TLU, then the response x_i can classify the input into two categoreis which are separated by the decision boundary defined by the discriminant function. A two-category classifier is termed a *pattern dichotomizer*. Operation of (9.92) is shown in Fig. 9.3.

Since the discriminant function of (9.92) is a linear weighted sum of the input, it defines a linear decision boundary, a *hyperplane* in general, and can thus correctly classify only *linearly separable* input vectors. More importantly, it was shown by Minsky and Pappert in their influential book *Perceptrons* (1969) that single feed forward slabs of TLU's, as in Rosenblatt's perceptron (1962), can only perform linearly separable

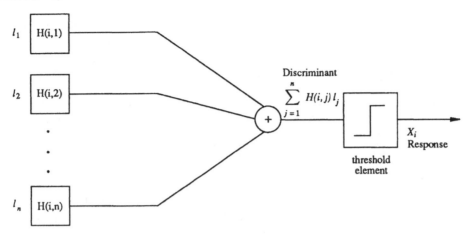

Figure 9.3 The threshold logic unit as a pattern classifier.

mappings. It was also suggested in that book that cascades of slabs of first order units will be of no more practical interest than single layers. The latter was an incorrect generalization as proven by the immense popularity of error backpropagation algorithm (Rumelhart et al. 1986) which trains multilayer networks to classify both linearly- and nonlinearly-seperable input patterns. This linear separability constraint is often cited as one of the most important reason for the previous decline of interest in neural network theory and application.

Inspection of Eq. (9.91), which is an expansion of the simplest shunting network with only inhibitory connections, shows that the term inside the brackets is an nth order discriminant function and hence a shunting net can perform an nth order classification. This result is interesting especially in light of the fact that only one transistor per interconnection is necessary for implementation of such a network.

The limitations of one-layer shunting nets for classification should also be expressed. The decision boundaries defined by (9.91) are rigid since the weights are constant. More importantly, the higher order weights are not independent in the Volterra expansion case and can not be individually set. This limits the shape of the decision boundaries and hence the type of nonlinearly separable functions that can be correctly classified by this class of one-layer neural networks. The mean input intensity dependence of the kernels, however, can be used to alter, and hence control, the shape of the decision regions. This may extent the results from the analysis of neural networks to their synthesis.

Units of the form (9.91) have been termed high-order neurons (Maxwell et al. 1986, Giles and Maxwell 1987) and have been shown to possess very desirable learning, generalization, and knowledge representation capabilities and can be hand crafted to solve many interesting high order problems.

Finally, Minsky and Pappert were well aware of high order discriminant functions but correctly observed that implementation of such units would suffer from combinatorial explosion of weights. This treatment shows that all high order weights may be implicit in the closed form of a simple network with only multiplicative inhibition.

9.6 CONCLUSIONS

The Volterra-Weiner series expansion can be applied to one-layer neural network models to allow for systematic inclusion of higher-order nonlinearities in the description of temporal and steady-state behavior of the network. This technique was shown to be especially suitable for modeling the adaptation of temporal response and the spatial receptive field of the networks to variation in mean input levels.

In this chapter, temporal and spatial Volterra kernels for a neural network model whose activity is governed by multiplicative lateral inhibition were derived. The spatial expansion showed that a network with first order weights can exhibit higher order properties such as classification of non-linearly separable input functions. Formulas derived for high order kernels (weights) show their dependence on lower order terms, and ultimately on mean value of the input. Since these weights define the shape of the decision regions, this limits the functions that can be classified but also shows that changing the mean input (or the total activity, or the background or ambient light in an optical input case) is one mechanism through which some control over the decision regions can be exerted.

The mathematical analysis, relation to vision science, neural network theory, image processing, motion detection and technological applications discussed in this book all have a common base, namely the study of the global behavior arising from local activity of simple units. The final conclusion is so simple as to be redundant: A number of simple units, be it neurons or transistors, if connected correctly, can perform *very* complicated tasks.

Bibliography

Ackley, D. H.; Hinton, G. E.; and Sejnowski, T. J. 1985. A learning algorithm for Boltzman machines. *Cog. Sci.* **9**: 147–169.

Adelson, E. H., and Bergen, J. R. 1985. Spatiotemporal energy models for the perception of motion. *J. Opt. Soc. Am.* **A2**: 284–299.

Agar, W. O., and Blythe, J. H. 1968. An optical method of measuring transverse surface velocity. *J. Scient. Instrum. (J. Physics E) Ser. 2* **1**: 25–28.

Alspector, J., and Allen, R. B. 1987. A neuromorphic VLSI learning system. *Proc. Stanford Conf. Advanced Research in VLSI* 313–349. Cambridge: MIT Press.

Alspector, J.; Allen, R. B.; Hu, V.; and Satyanarayana, S. 1987. Stochastic learning networks and their electronic implementation. *IEEE Conf. Neural Info. Proc. Sys.: Natural and Synthetic,* Denver.

Anderson, J. A. 1983. Cognitive and psychological computation with neural models. *IEEE Trans. Sys., Man, Cybern.* **13** (5): 799–815.

Anderson, J. A., and Rosenfeld, E. 1988. *Neurocomputing: Foundations of Research.* Cambridge: MIT Press.

Anderson, J. A.; Silverstein, J. W.; Ritz, S. R.; and Jones, R. S. 1977. Distinctive features, categorical perception, and probility learning: Some application of neural models. *Psy. Rev.* **84**: 413–451.

Anstis, S. M. 1980. The perception of apparent movement. *Phil. Trans. Roy. Soc. Lond.* **B290**: 153–168.

Ariel, M., and Adolph, A. R. 1985. Neurotransmitter inputs to directionally sensitive turtle ganglion cells. *J. Neurophysiol.* **54:** 1123–1143.

Ariel, M., and Daw, N. W. 1982. Pharmacological analysis of directionally sensitive rabbit retinal ganglion cells. *J. Physiol. Lond.* **324:** 161–185.

Ator, J. T. 1963. Image velocity sensing with parallel-slit reticles. *J. Opt. Soc. Am.* **53:** 1416–1422.

———. 1966. Image velocity sensing by optical correlation. *Appl. Opt.* **5:** 1325–1331.

Barlow, H. B. 1962. Three points about lateral inhibition. *Sensory Comm.* W. A. Rosenblith, ed. Cambridge: MIT Press.

———. 1965. Optic nerve impulses and Weber's law. *Cold Spring Harbor Sym. on Quant. Biol.* 539–546.

———. 1981. Critical limiting factors in the design of the eye and visual cortex. *Proc. Roy. Soc. Lond.* **B212:** 1–34.

Barlow, H. B., and Levick, W. R. 1965. The mechanism of directionally selective units in the rabbit's retina. *J. Physiol.* **178:** 477–504.

Barlow, H. B.; Hill, R. M.; and Levick, W. R. 1964. Retinal ganglion cells responding selectively to direction and speed of image motion in the rabbit. *J. Physiol.* **173:** 377–407; 477–504.

Barnard, S. T., and Thompson, W. B. 1980. Disparity analysis of images. *IEEE Trans. Pattern Analy. Machine Intell.* **PAMI-2:** 333–340.

Baylor, D. A.; Hodgkin, A. L.; and Lamb, T. D. 1974. Reconstruction of electrical responses of turtle cones to flashes and steps of light. *J. Physiol.* **224:** 759–791.

Bell, H. H., and Lappin, J. S. 1973. Sufficient conditions for the discrimination of motion. *Perception and Psychophysics* **14:** 45–50.

Beverley, K. I., and Regan, D. 1973. Evidence for the existence of neural mechanisms selectively sensitive to the direction of movement in space. *J. Physiol.* **235:** 17–29.

Bishop, L. G., and Keehn, D. G. 1967. Neural correlates of the optomotor response in the fly. *Kybernetik* **3:** 288–295.

Bishop, L. G.; Keehn, D. G.; and McCann, G. D. 1968. Studies on motion detection by interneurons of the optic lobes and brain of the flies *Calliphora phaenicia* and *Musca domestica. J. Neurophysiol.* **31:** 509–525.

Blake, A., and Zisserman, A. 1987. *Visual Reconstruction.* Cambridge: MIT Press.

Blakemore, J. S. 1982. Semiconducting and major properties of gallium arsenide. *Multitopic Reviews.* 3–61.

Borst, A., and Egelhaaf, M. 1987. Temporal modulation of luminance adapts time constant of fly movement detectors. *Biol. Cybern.* **56:** 209–215.

Bouzerdoum, A., and Pinter, R. B. 1989. Biophysical basis, stability, and directional response characteristics of multiplicative lateral inhibitory neural networks. *Proc. Int. Joint Conf. Neural Net.* Washington D. C., June.

———. 1989a. Image motion processing in biological and computer vision systems. *Visual Comm. and Image Process. IV, Proc.* **SPIE 1199:** 1229–1240.

————. 1989b. Biophysical basis, stability, and directional response characteristics of multiplicative lateral inhibitory neural networks. *Int. Joint Conf. Neural Nets.* 18–22, Washington, D. C., June.

————. 1990a. A shunting inhibitory motion detector that can account for the functional characteristics of fly motion-sensitive interneurons, Proc. Int. Joint Conf. Neural Nets, 17–21 June 1990, San Diego, CA. I-149-I-153.

————. 1990b. Adaptation of directional response characteristics of a class of shunting inhibitory neural networks. *Nonlinear Met. of Physiol. Sys. Mod., BMSR workshop.* Los Angeles, CA., March.

Boycott, B. B., and Dowling, J. E. 1969. Organization of the primate retina: Light microscopy. *Phil. Trans. Roy. Soc.* **B255:** 109–176.

Boyd, S., and Chua, L. O. 1985. Fading memory and the problem of approximating Non-linear operators with Volterra series. *IEEE Trans. on Circ. and Sys.* **32:** 1150–1160.

Braddick, O. J. 1974. A short-range process in apparent motion. *Vision Res.* **14:** 519–527.

————. 1980. Low level and high level processes in apparent motion. *Phil. Trans. Roy. Soc. Lond.* **B290:** 137–151.

Buchner, E. 1976. Elementary movement detectors in an insect visual system. *Biol. Cybern.* **24:** 85–101.

Bülthoff, H.; Little, J.; and Poggio, T. 1989. A parallel algorithm for real-time computation of optical flow. *Nature* **337:** 549–555.

Caldwell, J. H.; Daw, N. W.; and Wyatt, H. J. 1978. Effects of picrotoxin and strychnine on rabbit retinal ganglion cells: Lateral interactions for cells with more complex receptive fields. *J. Physiol. Lond.* **276:** 277–298.

Card, H. C., and Moore, W. R. 1989. VLSI devices and circuits for neural networks. *Int. J. Neural Sys.* **1** (2): 149–165.

Carpenter, G. A., and Grossberg, S. 1987a. A massively parallel architecture for a self-organizing neural pattern recognition machine. *Computer Vision, Graphics, and Image Processing* **37**: 54–115.

————. 1987b. ART 2: Self organization of stable category recognition codes for analog input patterns. *Appl. Opt.* **26:** 4919–4930.

————. 1988. The ART of adaptive pattern recognition by self-organizing neural networks. *Computer* **21:** 77–88.

Chen, H. W.; Jacobson, L. D.; Gaska, J. P.; and Pollen, D. A. 1989. Structural classification of multi-input biological nonlinear systems. *Proc. IEEE Conf. Sys., Man and Cybern.* **IEEE-SMC:** 903–908.

Clocksin, W. F. 1980. Perception of surface slant and edge labels from optical flow: A computational approach. *Perception* **9:** 253–269.

Cohen, M. A. 1988. Sustained oscillations in a symmetric cooperative-cometetive neural network: Disproof of a conjecture about content addressable memory. *Neural Networks* **1:** 217–221.

Cohen, M. A., and Grossberg, S. 1983. Absolute stability of global pattern formation and parallel memory storage by competitive neural networks. *IEEE Trans. Sys., Man, Cybern.* **SMC-13:** 815–826.

———. 1984. Neural dynamics of brightness perception: Features, boundaries, diffusion, and resonance. *Perception and Psychophysics* **36 (5):** 428–456.

Cornsweet, T. N. 1970. *Visual Perception* New York: Academic Press.

Cornsweet, T. N., and Yellott, J. I., Jr. 1985. Intensity-dependent spatial summation. *J. Opt. Soc. Am.* **A2:** 1769–1786.

Curlander, J. C., and Marmarelis, V. Z. 1983. Processing of visual information in the distal neurons of the verteberate retina. *IEEE Trans. Sys., Man, Cybern.* **SMC-13:** 934–943.

———. 1987. A linear spatio-temporal model of the light-to-bipolar cell system and its response characteristics to moving bars. *Biol. Cybern.* **57:** 357–363.

Curtice, W. 1980. A MESFET model for use in the design of GaAs integrated circuits. *IEEE Trans. on Microwave Theory and Techniques* **MTT-28:** 448–456.

Czarnul, Z. 1986. Design of voltage-controlled linear transconductance elements with a matched pair of FET transistors. *IEEE Trans. Cir. Sys.* **33(10):** 1012–1015.

Daitch, J. M., and Green, D. G. 1969. Contrast sensitivity of the human peripheral retina. *Vision Res.* **9:** 947–952.

Darling, R. B., and Nabet, B. 1988. Prospects for integration of sensory neural networks in gallium arsenide photodetector arrays. *Northcon/88 Conf. Rec.* **II:** 946–952, Seattle, October.

Darling, R. B.; Nabet, B.; and Pinter, R. B. 1989. Implementation of analog shunting neural networks for optoelectronic detection and processing. *IEEE Proc. Int. Sym. Cir. Sys.* 465–469, Portland, OR, May.

Darling, R. B.; Nabet, B.; Samaras, J. E.; Ray, S.; and Carter, E. 1988. Epitaxial n$^+$ GaAs mesa-finger interdigital surface photodetectors. *IEEE Electron Device Lett.* **10(10):** 461–463.

J. J. D'Azzo and C. H. Houpis. 1978. *Feedback Control System Analysis and Synthesis.* New York: McGraw-Hill.

de Ruyter van Steveninck, R. R.; Zaagman, W. H.; and Mastebroek, H. A. K. 1986. Adaptation of transient responses of a movement-Sensitive neuron in the visual system of the blowfly *Calliphora erythrocephala. Biol. Cybern.* **54:** 223–236.

Delbrück, T., and Mead, C. A. 1989. An electronic photoreceptor sensitive to small changes in intensity. *Advances in Neural Information Processing Systems, 1.* D. S. Touretzky, ed. 720–727. San Mateo: Morgan Kaufmann.

Derrington, A. M., and Lennie, P. 1982. The influence of temporal frequency and adaptatiion level on receptive field organization of retinal ganglion cells in cat. *J. Physiol.* **333:** 343–366.

Dubs, A. 1982. The spatial integration of signals in the retina and lamina of the fly compound eye under different conditions of luminance. *J. Comp. Physiol.* **46:** 321–343.

Dvorak, D.; Srinivasan, M. V.; and French, A. S. 1980. The contrast sensitivity of fly movement-Detection neurons. *Vision Res.* **20**: 397–407.

Egelhaaf, M. 1985. On the neuronal basis of figure-ground discrimination by relative motion in the visual system of the fly: I. Behavioural constraints imposed on the neuronal network and the role of the optomotor system. *Biol. Cybern.* **52**: 123–140.

Egelhaaf, M., and Reichardt, W. 1987. Dynamic response properties of movement detectors: theoretical analysis and electrophysiological investigation in the visual system of the fly. *Biol. Cybern.* **56**: 69–87.

Ellias, S. A., and Grossberg, S. 1975. Pattern formation, contrast control, and oscillations in the short term memory of shunting on-center off-surround networks. *Biol. Cybern.* **20**: 69–98.

Emerson, R. C.; Citron, M. C.; Vaughn, W. J.; and Klein, S. A. 1987. Nonlinear directionally selective subunits in complex cells of cat striate cortex. *J. Neurophysiol.* **58**: 33–65.

Enroth-Cugell, C., and Robson, J. G. 1966. The contrast sensitivity of retinal ganglion cells of the cat. *J. Physiol.* **187**: 517–552.

Exner, S. 1875. Uber das Sehen von Bewegungen und die Theorie des Zusammengesetzten Auges. *Sitzungsber Akad Wiss Wien Abt III* **72**: 156–190.

Fahrenbach, W. 1985. The anatomical circuitry of lateral inhibition in the eye of the horseshoe crab, *Limulus polyphemus. Proc. Roy. Soc. Lond.* **B225**: 219–249.

Fennema, C. L., and Thompson, W. B. 1979. Velocity determination in scenes containing several moving images. *Comp. Graph. Image Process.* **9**: 301–315.

Fiorentini, A., and Maffei, L. 1973. Contrast in night vision. *Vision Res.* **13**: 73–80.

Foster, D. H. 1971. A model of the human visual system in its response to certain classes of moving stimuli. *Kybern.* **8**: 69–84.

Franceschini, N.; Riehle, A.; and Le Nestour, A. 1989. Directionally selective motion detection by insect neurons. *Facets of Vision*, D. G. Stavenga and R. C. Hardie, eds. 360–389. Springer Verlag: Berlin.

Franklin, J. N. 1968. *Matrix Theory.* Englewood Cliffs: Prentice-Hall.

Fraser-Rowell, C. H.; O'Shea, M.; and Williams, J. L. D. 1977. The neuronal basis of a sensory analyzer, the acridid movement detector system IV. The preference for small field stimuli. *J. Exp. Biol.* **68**: 157–185.

Freeman, W. J. 1979. *Mass Action in the Nervous System.* New York: Academic Press.

———. 1979. Nonlinear dynamics of paleocortex manifested in the olfactory EEG. *Biol. Cybern.* **35**: 21–37.

Frisby, J. P. 1972. The effect of stimulus orientation on the Phi phenomenon. *Vis. Res.* **12**: 1145–1162.

Fu, C. W., and Chang, S. 1989. A motion estimation algorithm under time-varying illumination. *Pattern Recog. Letters* **10**: 195–199.

Fuortes, M. G. F., and Hodgkin, A. L. 1964. Changes in time scale and sensitivity in the ommatidia of *Limulus. J. Physiol.* **172**: 239–263.

Fuortes, M. G. F., and Yeandle, S. 1964. Probability of occurrence of discrete potential waves in the eye of *Limulus. J. Gen. Physiol.* **47:** 443–463.

Furman, G. G. 1965. Comparison of models for subtractive and shunting lateral inhibition in receptor-neuron fields. *Kybernetic* **2:** 257–274.

Furman, G. G., and Frischkopf, L. S. 1964. Model of neural inhibition in the mammalian cochlea. *J. Acoust. Soc. Am.* **36:** 2194–2201.

Gennery, D. B. 1979. Object detection and measurement using stereo vision. *Proc. 6th Int. Joint Conf. Artif. Intell.* 320–327. Tokyo.

Gerstein, G. L.; Bedenbaugh, P.; and Aertsen, A. M. H. J. 1989. Neuronal assemblies. *IEEE Trans. Biomed. Eng.* **36 (1):** 4–14.

Giles, C. L.; Maxwell, T. 1987. Learning invariance and generalization in high-order neural networks. *App. Opt.* **26 (23):** 4972–4977.

Graf, H. P., and deVegver, P. 1987. A CMOS associative memory chip based on neural networks. *Digest ISSCC* **304:** New York.

———. 1987. A CMOS implementation of a neural network model. *Proc. Stanford Conf. Advanced Research in VLSI* 352–367. Cambridge: MIT Press.

Graf, H. P.; Jackel, L. D.; Howard, R. E.; Straughn, B.; Denker, J. S.; Hubbard, W.; Tennant, D. M.; and Shwartz, D. 1986. VLSI implementation of a neural network with several hundreds of neurons. *Proc. Conf. Neural Networks for Computing,* Denker, J. S., ed. Snowbird, Utah. American Inst. of Physics.

Grossberg, S. 1968. Some nonlinear networks capable of learning a spatial pattern of arbitrary comlexity. *Proc. Nat. A. of Sci.* **59:** 368–372.

———. 1969. Some networks that can learn, remember, and reproduce any number of complicated space-time patterns, I. *J. Math. and Mech.* **19:** 53–91.

———. 1970. Some networks that can learn, remember, and reproduce any number of complicated space-time patterns, II. *Stu. App. Math.* **49:** 135–166.

———. 1973. Contour enhancement, short term memory and constancies in reverberating neural networks. *Stu. App. Math.* **52:** 217–257.

———. 1976. Adaptive pattern classification and universal recoding, I: Parallel development and coding of neural feature detectors. *Biol. Cybern.* **23:** 121–134.

———, ed. 1981. Adaptive resonance in development, perception, and cognition. *Math. Psy. and Psychophys.* Providence, RI: American Mathematical Society.

———. 1982. *Studies of Mind and Brain: Neural Principals of Learning, Perception, Development, Cognition, and Motor Control.* Boston: Reidel Press.

———. 1983. The quantized theory of visual space: The coherent computaion of depth, form, and lightness. *Behav. and Brain Sci.* **6:** 625–692.

———, ed. 1986a. *The Adaptive Brain I: Cognition, Learning, Reinforcement and Rhythm,* Amsterdam: Elsevier/North-Holland.

———,ed. 1986b. *The Adaptive Brain II: Vision, Speech, Language and Motor Control,* Amsterdam: Elsevier/North-Holland.

———. 1987. Competitive learning: From interactive activation to adaptive resonance. *Cog. Sci.* **11**: 23–63.

———. 1988a. Non-linear neural networks: principles, mechanisms, and architectures. *Neural Networks,* **1**: 17–61.

———. ed. 1988b. *Neural Networks and Natural Intelligence*

Grossberg, S., and Levine, D. 1975. Some developmental and attentional biases in the contrast enhancement and short term memory of recurrent neural networks. *J. Theor. Biol.* **53**: 341–380.

Grossberg, S., and Mingola, E. 1985. Neural dynamics of perceptual grouping: Textures, boudaries, and emergent segmentations. *Percep. and Psychophys.* **38** (2) 141–171.

Grossberg, S., and Mingola, E. 1987. Neural dynamics of surface perception: Boundary webs, illuminants, and shape-from-shading. *Comp. Vis. Grap. and Image Proc.* **37**: 116–165.

Grossberg, S., and Rudd, M. E. 1989. A neural architecture for visual motion perception: Group and element apparent motion. *Neural Net.* **2**: 421–450.

Grossberg, S., and Todorovic, D. 1988. Neural dynamics of 1-D and 2-D brightness perception: A unified model of classical and recent phenomena. *Percep. and Psychophys.* **43**: 241–277.

Grossberg, S.; Mingolla, E.; and Todorovic, D. 1989. A neural network architecture for preattentive vision. *IEEE Trans. Biomed. Eng.* **36** (1): 65–84.

Harris, M. G. 1986. The perception of moving stimuli: A model of spatiotemporal coding in human vision. *Vis. Res.* **26**: 1281–1287.

Hartline, H. K. 1940. The receptive field of optic nerve fibers. *Am. J. Physi.* **130**: 690–699

Hartline, H. K., and Ratliff, F. 1974. *Studies in Excitation and Inhibition in the Retina.* New York: Rockefeller University Press.

Hassenstein, B., and Reichardt, W. 1956. Functional structure of a mechanism of perception of optical movement. *Proc. I. Int. Cong. Cybern. Namur.* 797–801.

Hausen, K. 1981. Monocular and binocular computation of motion in the lobula plate of the fly. *Dtsch. Zool. Ges.* **74**: 49–70.

Hebb, D. O. 1949. *The Organization of Behavior.* New York: Wiley.

Heeger, D. J. 1988. Optical flow from spatiotemporal filters. *Int. J. Comp. Vis.* **1**: 279–302. Also, *Proc. 1st Int. Conf. Comp. Vis.* London, 181–190.

Hildreth, E. C. 1984. The Computation of the Velocity Field. *Proc. Roy. Soc. Lond.* B221: 189–220.

Hodgkin, A. L., and Huxley, A. F. 1952. A quantitative description of membrane current and its application to conduction and excitation in nerve. *J. Physio.* **117** 500–544.

Hopfield, J. J. 1982. Neuronal networks and physical systems with emergent collective computational abilities. *Proc. Nat. A. Sci.* **79**: 2554–2558.

———. 1984. Neurons with graded response have collective computational properties like those of two state neurons. *Proc. Nat. A. Sci.* **81**: 3058–3092.

———. 1990. The effectiveness of analogue neural network hardware. *Network* **1**: 27–40.

Horn, B. K. P. 1987. Motion fields are hardly ambiguous. *Int. J. Comp. Vision,* **1**: 259–274.

Horn, B. K. P., and Schunck, B. G. 1981. Determining optical flow. *Artificial Intell.* **17**: 185–203.

Horowitz, P., and Hill, W. 1980. *The Art of Electronics*. Cambridge: Cambridge University Press.

Huang, T. S. 1981. *Image Sequence Analysis*. New York: Springer Verlag.

Hubel, D. H., and Weisel, T. N. 1959. Receptive fields of single neurons in the cat's striate cortex. *J. Physiol.* **148**: 574–591.

———. 1962. Receptive fields, binocular interaction and functional architecture in the cat's visual cortex. *J. Physiol.* **160**: 106–154.

Jacobus, C. J.; Chien, R. T.; and Selander, J. M. 1980. Motion detection and analysis by matching graphs of intermediate-level primitives. *IEEE Trans. Pattern Analy. Machine Intell.* **PAMI-2**: 495–510.

Jain, R.; Martin, W. N.; and Aggarwal, J. K. 1979. Segmentation through the detection of changes due to motion. *Comp. Graph. Image Process.* **11**: 13–34.

Jain, R.; Militzer, D.; and Nagel, H. H. 1977. Separating nonstationary from stationary scene components in a sequence of real-world TV images. *Proc. 5th Int. Joint Conf. Artif. Intell.* 612.

Jain, R., and Nagel, H. H. 1979. On the analysis of accumulative difference picture from image sequences of real world scenes. *IEEE Trans. Pattern Analy. Machine Intell.* **PAMI-1**: 206–214.

Jernigan, M. E., and Wardell, R. W.; 1981. Does the eye contain optimal edge detection mechanisms?. *IEEE Trans. Sys., Man, and Cybern.* **SMC-11**: 441–444.

Jernigan, M. E.; Belshaw, R. J.; and McLean, G. F. 1989. Image enhancement with nonlinear local interaction. *IEEE Proc. Int. Con. Sys., Man, and Cybern.* **IEEE-SMC**: 676–681.

Julesz, B. 1971. *Foundations of Cyclopean Motion*. Chicago: University of Chicago Press.

Kandel, E. R. 1976. *Cellular Basis of Behavior*. San Fransisco: Freeman and Company.

Kearney, J. K.; Thompson, W. B.; and Boley, D. L. 1987. Optical flow estimation: An error analysis of gradient-based methods with local optimization. *IEEE Trans. Pattern Analy. Machine Intell.* **PAMI-9**: 229–244.

Kelly, D. H. 1975. Spatial frequency selectivity in the retina. *Vision Res.* **15**: 665–672.

Koch, C., and Segev, I., eds. 1989. *Methods in Neuronal Modeling: From Synapses to Networks*. Cambridge: MIT Press.

Koch, C.; Poggio, T.; and Torre, V. 1982. Retinal ganglion cells: A functional interpretation of dendritic morphology. *Phil. Trans. Roy. Soc. Lond.* **B 298**: 227–264.

Kohonen, T. 1977. *Associative Memory—A System Theoretical Approach*. New York: Springer Verlag.

————. 1984. *Self-Organization and Associative Memory* New York: Springer Verlag.

————. 1988. An introduction to neural computing. *Neural Network* **1**: 3–16

Kohonen, T., and Oja, E. 1976. Fast adaptive formation of orthogonalizing filters and associative memory in recurrent networks of neuron-like elements. *Bio. Cybern.* **21**: 85–95.

Kolers, P. A. 1972. *Aspects of Motion Perception.* New York: Pergamon Press.

Kosko, B. 1987. *Bidirectional Associative Memory*

Kuffler, S. W. 1953. Discharge pattern and functional organization of the retina. *J. Neurophysiol.* **16**: 37–68.

Kuffler, S. W., and Nicholas, J. G. 1976. *From Neuron to Brain.* Massachussettes: Sinauer Associates.

Kuno, M., and Miyahara, J. T. 1969. *J. Physiol.* **201**: 465.

Lai, S. H., and Chang, S. 1988. Estimation of 3-D translational motion parameters via Hadamard transform. *Pattern Recog. Letters* **8**: 341–345.

Laughlin, S. B. 1983. Matching coding to scenes to enhance efficiency. *Physical and Biological Processing of Images,* Braddick, O. J. and Sleigh, eds. 42–52, Berlin: Springer Verlag.

————. 1989. The role of sensory adaptation in the retina. *J. exp. Biol.* **146**: 39–62.

Lawton, D. 1983. Processing translational motion sequences. *Comp. Vision, Graph. Image Process.* **22**: 116–144.

Lazzaro, J.; Ryckebusch, S.; Mahowald, M. A.; and Mead, C. A. 1989. Winner-take-all networks of O(N) complexity. *Advances in Neural Information Processing Systems,* D. S. Touretzky, ed. **1**: 703–711. San Mateo: Morgan Kaufmann.

Lee, Y. W., and Schetzen, M. 1965. Measurement of the Wiener kernels of a non-linear system by cross correlation. *Int. J. Cont.* **2**: 237–254.

Leese, J. A.; Novak, C. S.; and Taylor, V. R. 1970. The determination of cloud pattern motion from geosynchronous satellite image data. *Pattern Recog.* **2**: 279–292.

Lehovec, K., and Zuleeg, R. 1970. Voltage-current characteristics of GaAs J-FETs in the hot electron range. *Solid-State Elec.* **10**: 1415–1426.

Lettvin, J. Y.; Maturana, H. R.; McCulloch, W. S.; and Pitts, W. H. 1959. What the frog's eye tells the frog's brain. *Proc. IRE,* **47**: 1940–1951.

Levine, D. S. 1983. Neural population modelling and psychology: A review. *Math. Biosciences* **66**: 1–86.

Levine, M. D. 1985. *Vision in Man and Machine* New York: McGraw Hill.

Lippmann, R. P. 1987. An introduction to computing with neural nets. *IEEE ASSP Magazine* **4**: 4–22.

Longuet-Higgins, H. C., and Prazdny, K. 1980. The interpretation of moving retinal images. *Proc. Roy. Soc. Lond.* **B208**: 385–397.

Lotka, A. J. 1956. *Elements of Mathematical Biology.* New York: Dover.

Mackie, S.; Graf, H. P.; and Shwartz, D. B. 1988. Implementation of neural network models in silicon. *Neural Computers*, Eckmiller, R., and Malsburg, V. D. C., eds. Heidelburg: Springer Verlag.

Marmarelis, P. Z., and Marmarelis, V. Z. 1978. *Analysis of Physiological Systems*. New York: Plenum Press.

Marmarelis, P. Z., and McCann, G. D. 1973. Development and application of white-noise modeling techniques for studies of insect visual nervous system. *Kybernetik* **12:** 74–90.

Marr, D. C. 1982. *Vision: A Computational Investigation Into the Human Representation and Processing of Visual Information*. San Fransisco: W. H. Freeman.

Marr, D. C., and Hildreth, E. 1980. Theory of edge detection. *Proc. Roy. Soc. Lond.* **B207:** 187–217.

Marr, D. C., and Poggio, T. 1976. Cooperative computation of stereo disparity. *Science* **194:** 283–287.

Marr, D. C., and Ullman, S. 1981. Direction selectivity and its use in early visual processing. *Proc. Roy. Soc. Lond.* **B211:** 151–180.

Marshall, J. A. 1990. Self-organizing neural networks for perception of visual motion. *Neural Networks* **3:** 45–74.

Martin, W. N., and Aggarwal, J. K. 1979. Dynamic scene analysis: The study of moving images. *Inf. Syst. Res. Lab, E. Res. Center* **TR-184**.

Maturana, H. R., and Frenk, S. 1963. Directional movement and horizontal edge detectors in the pigeon eye. *Science* **142:** 977–979.

Maxwell, T.; Giles, C. L.; Lee, Y. C.; and Chen, H. H. 1986. Nonlinear dynamics of artificial neural systems.

McCulloch, W. S. 1965. *Embodiments of Mind*. Cambridge: MIT Press.

McCulloch, W. S., and Pitts, W. 1943. A logical calculus of the ideas immenant in nervous activity. *B. Math. Biophys.* **5:** 115–133.

McLean, G. F.; Jernigan, M. E.; St.Martin, I.; and Schriber, L. 1988. Adaptive image processing using nonlinear inhibition. *IEEE SMC Conf. Proc.* 563–566.

Mead, C. A. 1985. A sensitive electronic photoreceptor. *Chapel Hill Conference on VLSI* 463–471.

———. 1989. *Analog VLSI and Neural Systems*. Reading, Massachusetts: Addison-Wesley.

Mead, C. A., and Mahowald, M. A. 1988. A silicon model of early visual processing. *Neural Networks*, **1:** 91–97.

Merckel, G.; Borel, J.; and Cupcea, N. Z. 1972. An accurate large-signal MOS transistor model for use in computer-aided design. *IEEE Trans. Electron Dev.* **ED-19:** 681–690.

Michael, C. R. 1968. Receptive fields of single optic nerve fibers in a mammal with an all-cone retina: II. Directionally sensitive units. *J. Neurophysiol.* **31:** 257–2617.

Minsky, M. L. 1967. *Computation: Finite and Infinite Machines*. Englewood Cliffs: Prentice-Hall.

Minsky, M. L., and Pappert, S. 1969. *Perceptrons*. Cambridge: MIT Press.

Moller, A. R. 1987. Analysis of the auditory system using pseudorandom noise. *Advanced Methods of Physiological System Modeling I*. V. Z. Marmarelis, ed. Los Angeles: Biomedical Simulations Resource, Univ. of Southern California.

Moopenn, A.; Lambe, J.; and Thakoor, A. P.; 1987. Electronic implementation of associative memory based on neural network models. *IEEE Trans. on Sys., Man, and Cybern.* **SMC-17 (2):** 325–331.

Movshon, J. A.; Thompson, I. D.; Tolhurst, D. J. 1978. Receptive field organization of complex cells in the cat's striate cortex. *J. Physiol.* **283:** 79–99.

Murray, A. F.; Hamilton, A.; and Tarassenko L. 1989. Programmable analog pulse-firing neural networks. D. S. Touretzky, ed. *Advances in neural information processing systems* **1:** 671–677. San Mateo: Morgan Kaufmann.

Nabet, B. 1988. An introduction to adaptive resonance theory. *An Introduction to Artificial Neural Systems—A Video Short Course*, R. J. Marks and L. E. Atlas, eds. Seattle: TIE program, Univ. of Washington.

———. 1989. Design, Analysis, and Gallium Arsenide Implementation of a Class of Neural Networks with Application to Vision and Optoelectronics. Ph.D. dissertation, Dept. of Electrical Engineering. Seattle: University of Washington.

———. 1990. *Volterra Series Expansion and One-Layer Neural Networks*. Submitted for publication.

Nabet, B., and Darling, R. B. 1988. Implementation of optical sensory neural networks with simple discrete and monolithic circuits (Abstract). *Neural Networks* **1 (1):** 396. Boston, MA.

Nabet, B.; Darling, R. B.; and Pinter R. B.; 1989. Analog implementation of shunting neural networks. *Advances in Neural Information Processing Systems*, D. S. Touretzky, ed. 695–702. San Mateo: Morgan Kaufmann.

———. 1990. *Implementation of Front-End Processor Neural Networks*. Submitted for publication.

Nagel, H. H. 1978. Formation of an object concept by analysis of systematic time variation in the optically perceptible environment. *Comp. Graph. Image Process.* **7:** 149–194.

———. 1983. Displacement vectors derived from second order intensity variations in image sequences. *Comp. Vision, Graph. Image Process.* **21:** 85–117.

———. 1986. Image sequences—ten (octal) years—from phenomenology towards a theoretical foundation. *Proc. 8th Int. Conf. Pattern Recog.* Paris, 1174–1185.

———. 1987. On the estimation of optical flow: Relations between different approaches and some new results. *Artificial Intell.* **33:** 299–324.

Nakayama, K. 1985. Biological image motion processing: A review. *Vision Res.* **25:** 625–660.

Nilsson, N. J. 1965. *Learning Machines*. New York: McGraw-Hill.

Ogura, H. 1986. Estimation of Wiener kernels of a nonlinear system and fast algorithm using digital Laguerre filters. *15th NIBB Conf. Inf. Process. Neuron Networks*, K. Naka and Y. Ando, eds. 14–63.

Ong, D. G. 1984. *Modern MOS Technology: Processes, Devices, & Design.* New York: McGraw-Hill.

Osorio, D.; Srinivasan, M. V.; and Pinter, R. B. 1990. What causes edge fixation in walking flies? *J. Exp. Biol.* **149:** 281–292.

Palka, J. 1967. An inhibitory process influencing visual responses in a fibre of the ventral nerve cord of locusts. *J. Insect Physiol.* **13:** 235–248.

Pankove, J. I. 1971. *Optical Processes in Semiconductors.* New York: Dover.

Patel, A. S. 1966. Spatial resolution by the human visual system. The effect of mean retinal illuminance. *J. Opt. Soc. Am.* **56:** 689–694.

Peli, T., and Lim, J. S. 1982. Adaptive filtering for image enhancement. *Opt. Eng.* **21:** 108–112.

Petersik, J. T. 1980. The effect of spatial and temporal factors on the perception of stroboscopic rotation simulations. *Perception* **9:** 271–283.

Petersik, J. T.; Hicks, K. I.; and Pantle, A. J. 1978. Apparent movement of successively generated subjective patterns. *Perception* **7:** 371–383.

Petrie, G. 1984. Practical implementation of nonlinear lateral inhibition using junction field-effect transistors. MSEE thesis. Seattle: University of Washington.

Pinter, R. B. 1966. Sinusoidal and delta function responses of visual cells of the *Limulus* eye. *J. Gen. Physiol.* **49:** 565–593.

———. 1977. Visual discrimination between small objects and large textured backgrounds. *Nature* **270:** 429–431.

———. 1979. Inhibition and excitation in the locust DCMD receptive field: spatial frequency, temporal and spatial characteristics. *J. Exp. Biol.* **80:** 191–216.

———. 1983a. Product term nonlinear lateral inhibition enhances visual selectivity for small objects or edges. *J. Theor. Biol.* **100:** 525–531.

———. 1983b. The electrophysiological bases for linear and for nonlinear product term lateral inhibition and the consequences for wide field textured stimuli. *J. Theor. Biol.* **105:** 233–243.

———. 1984. Adaptation of receptive field spatial organization by multiplicative lateral inhibition. *J. Theor. Biol.* **110:** 435–444.

———. 1985. Adaptation of spatial modulation transfer functions via nonlinear lateral inhibition. *Biol. Cybern.* **51:** 285–291.

———. 1987a. Kernel synthesis from non-linear multiplicative lateral inhibition. *Advanced methods of physiological system modelling*, V. Z. Marmarelis, ed. **1:** 258–277. Los Angeles: Biomedical Simulations Resource, Univ. of Southern California.

———. 1987b. Visual system neural networks: Feedback and feedforward lateral inhibition. *Systems and Control Encyclopedia*, M. G. Singh ed. 5060–5065. Oxford: Pergamon Press.

———. 1989. Rushtonian transformation as a function of nonlinear lateral inhibition. *Visual Neuroscience*, submitted.

Pinter, R. B., and Nabet, B. 1991. Field adaptation in visual systems is a function of nonlinear lateral inhibition. *Neural Networks III*, Antognetti, P. and Milutinovic, V., eds. Englewood Cliffs: Prentice-Hall.

Pinter R. B.; Osorio, D.; and Srinivasan, M. V. 1990. Shift of edge preference to scototaxis depends on mean luminance and is predicted by a matched filter hypothesis in fly lamina LMC cells. *Vis. Neurosci.* **4:** 579–584.

Plonsey, R., and Fleming, D. J. 1969. *Bioelectric Phenomena.* New York: McGraw-Hill.

Poggio, T., and Reichardt, W. 1973. Considerations on models of movement detection. *Kybernetik* **13:** 223–227.

Poggio, T., and Torre, V. 1978. A new approach to synaptic interactions. In Theoretical approaches to complex systems, *Lecture Notes in Biomathematics*, R. Heim and G. Palm eds. **21:** 89–115 Berlin, Heidelberg, and New York: Springer Verlag.

Potter, J. L. 1975. Velocity as a cue to segmentation. *IEEE Trans. Sys., Man, and Cybern.* **SMC-5:** 390–394.

Potter, J. L. 1977. Scene segmentation using motion information. *Comp. Graph. Image Process.* **6:** 558–581.

Prazdny, K. 1980. Egomotion and relative depth map from optical flow. *Biol. Cybern.* **36:** 87–102.

Rall, W. 1977. Core conductor theory and cable properties of neurons. *Handbook of Physiology: The Nervous System* E. R. Kandel, ed. **I (I):** 39–97. Bethesda, MD: American Physiological Society.

Ramachandran, V. S., and Anstis, S. M. 1986. The perception of apparent motion. *Scientific American* **254 (6):** 102–109.

Ratliff, F. 1965. *Mach Bands: Quantitative Studies on Neural Networks in the Retina.* New York: Holden-Day.

———. 1984. Why Mach bands are not seen at the edges of a step. *Vision Res.* **24:** 163–165.

Ratliff, F., and Hartline. 1959. The response of *Limulus* optic nerve fibers to patterns of illumination on the retinal mosaic. *J. Gen. Physiol.* **42:** 1241–1255.

Ratliff, F.; Knight, B. W.; Graham, N. 1969. On tuning and amplification by lateral inhibition. *Proc. U. S. Nat. Acad. Sci.* **62:** 733–740.

Reichardt, W. 1961. Autocorrelation, a principle for the evaluation of sensory information by the nervous system. *Sensory Communication*, W. A. Rosenblith, ed. Cambridge: MIT Press, 303–317.

———. 1986. Processing of optical information by the visual system of the fly. *Vision Res.* **26:** 113–126.

Reichardt, W., and Egelhaaf, M. 1988. Properties of individual movement detectors as derived from behavioural experiments on the visual system of the fly. *Biol. Cybern.* **58:** 287–294.

Reichardt, W., and Guo, A. 1986. Elementary pattern discrimination: Behavioural experiments with the fly *Musca Domestica. Biol. Cybern.* **53:** 285–306.

Reichardt, W., and Poggio, T. 1979. Figure-ground discrimination by relative movement in the visual system of the fly, Part I: Experimental results. *Biol. Cybern.* **35:** 81–100.

Reichardt, W.; Poggio, T.; and Hausen, K. 1983. Figure-ground discrimination by relative movement in the visual system of the fly, Part II: Towards the neural circuitry. *Biol. Cybern.* **46:** (Suppl.), 1–30.

Reichardt, W.; Schlögl, R. W.; and Egelhaaf, M. 1988. Movement detector of the correlation-type provide sufficient information for local computation of 2-D velocity field. *Naturwissenschaften* **75:** 313–315.

Reye, D. 1990. *Collision Avoidance During Flight in the Locust: a Role for the DCMD Neurons.* Submitted for publication.

Richter, J., and Ullman, S. 1982. A model for the temporal organization of X- and Y-type receptive fields in the primate retina. *Biol. Cybern.* **43:** 127–145.

———. 1986. Non-linearities in cortical simple cells and the possible detection of zero-Crossings. *Biol. Cybern.* **53:** 195–202.

Rosenblatt, F. 1958. The perceptron: A probabilistic model of information storage and organization in the brain. *Psych. Rev.* **65:** 386–408.

———. 1962. *Principles of Neurodynamics.* Spartan: New York.

Ruff, P. I.; Rauschecker, J. P.; and Palm, G. 1987. A model of direction-selective simple cells in the visual cortex based on inhibition asymmetry. *Biol. Cybern.* **57:** 147–157.

Rugh, W. J. 1981. *Non-linear System Theory—The Volterra-Wiener Approach.* Baltimore and London: John Hopkins University Press.

Rumelhart, D. E.; Hinton, G. E.; and Williams, R. J. 1986. Learning internal representation by error propagation. *Parallel Distributed Processing,* D. E. Rumelhart and J. L. McClelland, eds. Cambridge: MIT Press.

Rumelhart, D. E., and McClelland, J. L., eds. 1986. *Parallel Distributed Processing.* Cambridge: MIT Press.

Sakitt, B., and Barlow, H. B. 1982. A model for the economical encoding of the visual image in cerebral cortex. *Bio. Cybern.* **43:** 97–108.

Sakuranaga, M., and Naka, K. I. 1985. Signal transmission in the catfish retina; II transmission to type-n cell. *J. Neurophysiol.* **53:** 390–410.

Schetzen, M. 1980. *The Volterra and Wiener Theory of Non-linear Systems,* New York: John Wiley and Sons.

———. 1983. Non-linear system modelling based on the Wiener theory. *Proc. IEEE* **69** **(12):** 1557–1573.

Schiller, P. H.; Finlay B. L.; and Volman, S. F. 1976. Quantitative studies of single-cell properties in monkey striate cortex: I. Spatiotemporal organization of receptive field. *J. Neurophysiol.* **39:** 1288–1374.

Schmid, A., and Bülthoff, H. 1988. Using neuropharmacology to distinguish between excitatory and inhibitory movement detection mechanisms in the fly *Calliphora Erythrocephala*. *Biol. Cybern.* **59**: 71–80.

Schunck, B. G. 1985. Image flow fundamentals and future research. *Proc. IEEE Conf. Comp. Vision and Pattern Recog., San Francisco* 560–571.

———. 1986. The image flow constraint equation. *Comp. Vision, Graph. Image Process.* **35**: 20–46.

Sethi, I. K.; Salari, V.; and Vemuri, S. 1988. Feature-point matching in image sequences. *Pattern Recog. Letters* **7**: 113–121.

Shanmugam, K. S.; Dickey, F. M.; and Green, J. A.; An optimal frequency domain filter for edge detection in digital pictures. *IEEE Trans. on Pattern Anal. and Mach. Intell.* **PAMI-1 (1)**: 37–49, January 1979.

Shichman, H., and Hodges, D. A. 1968. Modeling and simulation of insulated gate field-effect transistor switching circuits. *IEEE J. Solid-State Cir.* SC-3: 285–289.

Siebert, W. McC. 1986. *Circuits, Signals and Systems.* Cambridge: MIT Press and New York: McGraw-Hill.

Sillito, A. M. 1975. The contribution of inhibitory mechanisms to the receptive field properties of neurones in the striate cortex of the cat. *J. Physiol.* **250** 305–329.

Sivilotti, M. A.; Mahowald, M. A.; and Mead, C. A. 1987. *Proc. Stanford Conf. Advanced Research in VLSI.* Cambridge: MIT Press.

Smith, E. A., and Phillips, D. R. 1972. Automated cloud tracking using precisely aligned digital ATS pictures. *IEEE Trans. Comp.* **C-21**: 715–729.

Srinivasan, M. V., and Dvorak, D. R. 1980. Spatial processing of visual information in the movement detecting pathway of the fly. *J. Comp. Physiol.* **140**: 1–23.

Srinivasan, M. V.; Laughlin, S. B.; and Dubs, A. 1982. Predictive coding: A fresh view of inhibition in the retina. *Proc. Roy. Soc. Lond.* **216** 427–459.

Srinivasan, M. V.; Pinter, R. B.; and Osorio, D. 1990. Matched filtering in the visual system of the fly: Large monopolar cells of the lamina are optimized to detect moving edges and blobs. *Proc. Roy. Soc. Lond.* **240**: 279–293.

Steinbuch, K., and Piske, U. A. W. 1963. Learning matrices and their applications. *Trans. IEEE Elect. Computers* **EC-12**: 846–862.

Stockham, T. G. 1972. Image processing in the context of a visual model. *Proc. IEEE* **60 (7)**: 828–841.

Sutter, E. E. 1987. A practical nonstochastic approach to nonlinear time domain analysis. *Advanced Methods of Physiological System Modeling I* V. Z. Marmarelis, ed. 303–315. Los Angeles: Biomedical Simulations Resource, Univ. of Southern California.

Sze, S. M. 1981. *Physics of Semiconductor Devices, 2nd edition.* New York: John Wiley.

Taylor, G. W. 1978. Subthreshold conduction in MOSFETs. *IEEE Trans. Electron Devices* **ED-25**: 337–350.

Taylor, J. G. 1990. A silicon model of vertebrate retinal processing. *Neural Networks* **3**: 171–178.

Thakoor, A. P. 1987. *Content-Addressable, High Density Memories Based on Neural Network Models.* Jet Propulsion Lab, Tech Report JPL D-4166.

Thakoor, A. P.; Lamb, J. L.; Moopenn, A.; and Lambe, J. 1986. Binary synaptic connections based on memory switching in a-Si:H. *Proc. Conf. Neural Networks for Computing*, Denker, J. S. ed. Snowbird, Utah: American Inst. of Physics. 426–431.

Thompson, W. B., and Barnard, S. T. 1981. Low-level estimation and interpretation of visual motion. *Computer* **14 (8):** 20–28.

Thorson, J. 1966. Small-signal analysis of a visual reflex in the locust: II. Frequency dependence. *Kybernetik* **3:** 53–66.

Torre, V., and Poggio, T. 1978. A synaptic mechanism possibly underlying directional selectivity to motion. *Proc. R. Soc. Lond.* **B. 202:** 409–416.

Toyoda, J. 1974. Frequency characteristics of retinal neurons in the carp. *J. Gen. Physiol.* **63:** 214–234.

Tretiak, O., and Pastor, L. 1984. Velocity estimation from image sequences with second order differential operators. *Proc. 7th Int. Conf. Pattern Recog.* Montreal, Que. 16–19.

Troy, J. B.; Einstein, G.; Shuurmans, R. P.; Robson, J. G.; Enroth-Cugell, C. H. 1989. Responses to sinusoidal gratings of two types of very nonlinear retinal ganglion cells of cat. *Visual Neuroscience* **3:** 213–233.

Tsividis, Y. P. 1987. *Operation and Modeling of the MOS Transistor.* New York: McGraw-Hill Book Company.

Tsividis, Y. P. and Satyanarayana, S. 1987. Analogue circuits for variable-synapse electronic neural networks. *Electronic Letters* **23 (24):** 1313–1314.

Ullman, S. 1978. Two dimensionality of the correspondence process in apparent motion. *Perception* **7:** 683–693.

———. 1979a. *The Interpretation of Visual Motion.* Cambridge: MIT Press.

———. 1979b. The interpretation of structure from motion. *Proc. Roy. Soc. Lond.* **B203:** 405–426.

———. 1980. The effect of similarity between line segments on the correspondence strength in apparent motion. *Perception* **9:** 617–626.

———. 1981. Analysis of Visual Motion by Biological and Computer Systems. *Computer* **14 (8):** 57–69.

Uras, S.; Girosi, F.; Verri, A.; and Torre, V. 1988. A computational approach to motion perception. *Biol. Cybern.* **60:** 79–87.

Uyemura, J. P. 1988. *Fundamentals of MOS Digital Integrated Circuits.* Reading, Massachusetts: Addison-Wesley.

van Doorn, A. J., and Koenderink, J. J. 1976. A directionally sensitive network. *Biol. Cybern.* **21:** 161–170.

———. 1983. The structure of the human motion detection system. *IEEE Trans. Sys., Man, and Cybern.* **SMC-13:** 916–922.

Van Overstraeten, R. J.; Declerck, G. J.; and Muls, P. A. 1975. Theory of MOS transistor in weak inversion—new method to determine the number of surface states. *IEEE Trans. on Electron Devices* **ED-22**: 282–288.

van Santen, J. P. H., and Sperling, G. 1984. Temporal covariance model of human motion perception. *J. Opt. Soc. Am.* **A1**: 451–473.

———. 1985. Elaborated reichardt detectors. *J. Opt. Soc. Am.* **A2**: 300–321.

Varju, D. 1962. Vergleich zweier modelle fur laterale inhibition. *Kybernetik* **1**: 200–208.

Verri, A., and Poggio, T. 1987. Against quantitative optical flow. *Proc. 1st Int. Conf. Comp. Vision* London 171–180.

von Bekesy, G. 1967. *Sensory Inhibition*. Princeton: Princeton University Press.

Watson, A. B., and Ahumada, A. J., Jr. 1985. Model of human visual-motion sensing. *J. Opt. Soc. Am.* **A2**: 322–342.

Wertheimer, M. 1912. Experimentelle studien uber das sehen von bewegen. *Z. Psychol.* **61**: 161–278. (Excerpted and translated in *Classics in Psychology,* ed. T. Shipley, New York: Philosophical Library 1961.)

Williams, T. D. 1980. Depth for camera motion in a real world-scene. *IEEE Trans. Pattern Analy. Machine Intell.* **PAMI-2**: 511–516.

Willshaw, D. J.; Buneman, O. P.; and Longuet-Higgins, H. C. 1969. *Nature* **222**: 960.

Wilson, H. R. 1985. A model for direction selectivity in threshold motion perception. *Biol. Cybern.* **51**: 213–222.

Wilson, H. R., and Bergen, J. R. 1979. A four mechanism model for threshold spatial vision. *Vision Research* **19**: 19–32, 1979.

Wolferts, K. 1974. Special problems in interactive image processing for traffic analysis. *Proc. 2nd Int. Joint Conf. Pattern Recog., Copenhagen* 1–2.

Wong, F., and Knight, B. W. 1980. Adapting-bump model for eccentric cells of *Limulus*. *J. Gen. Physiol.* **76**: 539–557.

Wright, G., and Jernigan, M. E. 1986. Texture discriminants from spatial frequency channels. *IEEE SMC Conference Proc.* 519–524.

Yachida, M.; Asada, M.; and Tsuji, S. 1981. Automatic analysis of moving images. *IEEE Trans. Pattern Analy. Machine Intell.* **PAMI-3**: 12–19.

Yakimovsky, Y., and Cunningham, R. 1978. A system for extracting three-dimensional measurements from a stereo pair of TV cameras. *Comp. Graph. Image Process.* **7**: 195–210.

Yariv, A. 1975. *Quantum Electronics, 2nd Edition*. New York: John Wiley.

Yasui, S.; Davis, W.; Naka, K. I. 1979. Spatio-temporal receptive field measurment of retinal neurons by random pattern stimulation and cross correlation. *IEEE Trans. Biomedical Engr.* **BME-26**: 263–272.

Zaagman, W. H.; Mastebroek, H. A. K.; and Kuiper, J. W. 1978. On the correlation model: performance of a movement detecting neural element in the fly visual system. *Biol. Cybern.* **31**: 163–168.

Index

T - #0665 - 101024 - C0 - 234/187/11 - PB - 9781138561816 - Gloss Lamination